LIVING WITH RISK

The geography of technological hazards

By

Susan L Cutter
Rutgers University

Edward Arnold
A division of Hodder & Stoughton
LONDON NEW YORK MELBOURNE AUCKLAND

First published in Great Britain 1993

Distributed in the USA by Routledge, Chapman and Hall Inc.
29 West 35th Street, New York, N1 10001

British Library Cataloguing in Publication Data

Cutter, Susan L.
Living with Risk : Geography of
Technological Hazards
I. Title
363.18

ISBN 0–340–52987–3

Library of Congress Cataloging-in-Publication Data

Cutter, Susan L.
Living with risk : the geography of technological hazards /
Susan L. Cutter.
p. cm.
Includes bibliographical references and index.
ISBN 0–340–52987–3 : $22.50
1. Technology—Risk assessment—Case studies.
2. Technology—Social aspects—Case studies. I. Title.
T174.5.C88 1993
363.1—dc20 93–3294
CIP

Typeset in Melior by Anneset, Weston-super-Mare, Avon.
Printed and bound in Great Britain for Edward Arnold, a division of Hodder and Stoughton Limited, Mill Road, Dunton Green, Sevenoaks, Kent TN13 2YA by Biddles Ltd, Guildford & King's Lynn.

Preface

The idea for this project originated with a conversation with Susan Sampford, then the Geography Editor at Edward Arnold, during a nice spring day in my office in New Brunswick. We were discussing emerging research areas where there was a dire need for introductory and advanced-text material. I mentioned my dismay at not having a hazards textbook or something that I could give to students that included a broader-based range of hazards than those originating from extreme natural events. Despite my protestations to the contrary, Susan persuaded me to think about writing such a book.

In the intervening years, sandwiched between academic life and motherhood, it became increasingly clear that just such a book was needed. There was an entire generation of students who knew very little about the interaction between society, technology, and the environment although they were extremely interested in these issues and moved by reports of planetary degradation.

This book is designed as an introduction into the risk and hazards literature, not an exhaustive coverage of it. I would not begin to suggest that we can cover all the hazards associated with nuclear technology in a simple chapter, likewise how societies and individuals perceive and respond to risks and hazards. I have simply selected a number of themes that run through the social sciences and explored them here. The case studies are designed to illustrate these themes in a more interesting and tangible way.

I have learned much during the writing of this book. I am especially grateful to colleagues who read various chapters with critical eyes: Frank Popper, Bill Solecki, Chip Clarke, and John Tiefenbacher. A special thanks goes to members of my Technological Hazards Seminar in the autumn of 1991, who were subjected to various drafts of the manuscript and provided extremely valuable feedback and data on transboundary wastes. I have also benefited from discussions with colleagues here at Rutgers including Caron Chess, Mike Greenberg, Dan Wartenberg, and Dee Garrison on many of the aspects of risk and hazards that are explored in this volume. A special debt of thanks goes to Brian J. L. Berry, William L. Thomas, and Gilbert F. White who have had major influences on my own intellectual development in the hazards field.

The production of the book would not have been realized without the gentle prodding of Laura McKelvie, the Geography Editor at Edward Arnold who sends such sweet faxes inquiring about the book's progress. The rest of the production team (Louise Thompson and Diane Leadbetter-Conway) were also extremely helpful. The artwork was done by Michael Siegel, Director of the Rutgers Computer Cartography Laboratory on a MacIntosh computer.

Last, but not least, families always suffer when books are being written and mine is no exception. Langdon, Nathaniel, and Megan all put up with an inattentive spouse and mother as the final push was made to get the manuscript done. They, better than anyone else, know what living with risk is all about.

Susan L Cutter
Milltown, New Jersey
October 1992

Contents

List of figures

List of tables

1

The nature and character of technological hazards

In the modern world, the most dangerous form of determinism is the technological phenomenon. It is not a question of getting rid of it, but, by an act of freedom, of transcending it (Ellul 1963, p. xxxiii).

Bhopal, Chernobyl, Love Canal, Seveso, Windscale. These toxic monuments are aprt of our modern vocabulary, for the mere mention of these names conveys images of death and lasting environmental damage. In addition to these well-known cases of technological failures, there are a myriad of less catastrophic events that result in environmental degradation and/or human injuries every day. These originate in many of our ordinary activities such as driving a car, painting our house, or eating. Pollution is one type of technological hazard, food additives, and pesticide residues are another. Large-scale industrial failures such as nuclear power plant accidents and chemical spills provide the most dramatic example of technological hazards.

Technological hazards arise from our individual and collective use of technology and present a very different set of problems and responses than natural hazards. Since technological hazards are often more pervasive and less publicly recognized than natural hazards, they also pose some unique management problems. For example, the public's response to technological hazards is often ambiguous, resulting in over-reactions, under-reactions, or no reactions. Unlike natural hazards, technological hazards often do not provide visual or auditory cues as to their onset. We cannot see, hear, smell, or taste the dangers of radiation; nor can we discern the dangers of a silent toxic cloud that wafts by. As a consequence, the public relies more on the scientific community and regulators for guidance in the hazard potential than is the case with earthquakes, floods, or hurricanes.

This chapter provides a general background on the nature, types, and character of technological hazards. There is an implicit assumption that geography matters, for it is the geography of technological hazards that permits us to understand the interaction between technology and society, the impacts of technology on society and the environment, and how society adjusts to or adapts to life under these hazardous conditions.

What are technological hazards?

Borrowing from the natural hazards literature, we can define technological hazards as the interaction between technology, society, and the environment. 'Risk' is the measure of likelihood of occurrence of the hazard. 'Hazard' is a much broader concept that incorporates the probability of the event happening, but also includes the impact or magnitude of the event on society and the environment, as well as the sociopolitical contexts within which these take place. Hazards are the threats to people and the things they value, whereas risks are measures of the threat of the hazards (Kates and Kasperson 1983; Whyte and Burton 1980).

Technological hazards are socially constructed, not acts of God or an extreme geophysical event; they are products of our society. As such, they are imbedded in larger political, economic, social, and historical contexts and are inseparable from them. It is impossible to understand the hazard without examining the context within which it occurs. The elements of complexity, surprise, and interdependence are governing characteristics of technological hazards.

The interaction of the risk and its context ultimately define the level of hazard as well as governing individual and collective responses. The distinction between the concepts of risk and hazard is important as it illustrates some of the underlying viewpoints on how we recognize and assess environmental threats (risks) and what we do about them (hazards). The divergence in viewpoints on how to identify, assess, and manage risks and hazards has led to different scientific approaches as well. Hazards management utilizes individual and collective strategies to reduce and mitigate the impacts of hazards on people and places. Risk assessment emphasizes the estimation and quantification of risk in order to determine acceptable levels of risk and safety; in other words to balance the risks of a technology or activity against its social benefits in order to determine its overall social acceptability. As we shall see later in this book, risk- assessment methods have severe limitations when applied to certain types of risks and as a cogent tool for public policy decision-making.

The need to differentiate between risk and hazards is most relevant in technological hazards because of the various ways that society governs and responds to the use and misuse of technology. The risks may always be present, but it is not until we have the interaction of the risks with people or places that they become hazards. Environmental or technological risk (I will use the term interchangeably), as a field of scientific inquiry is not solely the possession of the social scientist. Indeed, social scientists often rely on statisticians, engineers, and biomedical researchers to help estimate and ultimately quantify such risks. Hazards, however, are clearly within our realm of inquiry since we are interested in their location and distribution, the types of environments and people that might be exposed, the likely outcomes of such exposures, and finally the methods for lessening these human and environmental impacts.

Diversity and classification

One of the easiest ways to understand the range and diversity of technological hazards is by classifying them. Many authors have attempted such taxonomies like Starr's (1969) division based on involuntary or voluntary exposure, or Rowe's (1977) scheme based on general risk factors. Other researchers have provided different organizing themes to differentiate between types of technological hazards. Slovic (1987), for example, uses the public perception of risks to develop categories of risky technologies (see Chapter 2). Risks are classified based on two factors: dread and unknown. Risks that have a high dread factor and uncertain impacts such as nuclear-reactor accidents are at one extreme, while known risks with little or no dread (bicycles, power mowers) are at the other. Obviously, there are many technologies in between. Another approach is to use broad-based activities or technologies that cause or create environmental risks. These causal domains of environmental risk include the categories of public health, natural resources, economic development, disasters, and new products (Whyte and Burton 1980).

Still another classification scheme has evolved that places technological failures into two distinct categories: low-probability/high-consequence events (acute episodes or failures such as Three Mile Island), and high-probability/low-consequence events such as smog or water pollution. The latter events are distinguishable by a periodicity that enables some level of predictability based on occurrence, duration, and magnitude. The former events involve complex and tightly coupled systems that are renowned for the improbability of failure perversely creating what Perrow (1984) terms normal accidents.

One of the more sophisticated schemes was developed by Hohenemser, Kates and Slovic (1983) who describe a seven-class taxonomy based on 12 attributes or descriptors of the risk (Table 1.1). These attributes include the intent to harm, spatial extent of impact, concentration, persistence or length of time release remains a threat, recurrence or the time over which significant release recurs, population at risk, time between exposure and occurrence of consequences, annual human mortality, maximum human mortality, number of future generations at risk or transgenerational effects, potential non-human mortality, and actually experienced non-human mortality. Based on these descriptors, 93 events and activities were used to create a seven-class taxonomy. Multiple extreme hazards such as war and radiation are persistent in their effects and longevity, placing future generations at risk. Extreme hazards (further subdivided into five categories) have similar attributes such as a diffuse spatial impact (greenhouse warming), or are relatively common in their occurrence (automobile crashes). The last category, hazards, are distinguished by their lack of extreme characteristics. This grouping represents more of the everyday, relatively low-impact hazard such as food additives or appliance use (Table 1.1).

Table 1.1 Taxonomy of hazards.

Class	Example
1) Multiple extreme hazards	Nuclear war, radiation, nerve gas
2) Extreme hazards	
a) intentional biocides	Antibiotics, vaccines
b) persistent teratogens	Uranium mining, asbestosis
c) rare catastrophes	LNG explosions, airplane crashes
d) common killers	Car crashes, smoking, guns
e) diffuse global threats	Ozone depletion, greenhouse warming, AIDS
3) Hazards	Food additives, appliances

Source: Hohenemser, Kates and Slovic 1983, 1985.

These taxonomies help to place technology in perspective and serve as a useful compendium of the diversity of events that are considered technological hazards. All of these taxonomies illustrate a certain level of comparability that is possible between seemingly different events based on their hazard characteristics. What they do not provide is a ranking of the risks nor an assessment of which ones should receive our attention in order to reduce their consequences. As Hohenemser, Kates and Slovic (1985) remark, 'We regard neither triage nor adherence to cost-effectiveness criteria as adequate foundations for managing hazards; rather, we see them as the horns of a familiar dilemma: whether to work on the "big questions" where success is limited, or to work on the normal, where success is expected' (p. 85).

Geography matters

Geographic scale is important in understanding technological hazards, their distribution, impacts, and reduction. It is much easier to address a rather localized hazard source, such as a refinery spewing toxic chemicals from its smokestack than it is to address the cumulative impacts of greenhouse-gas emissions on global climate change. The level of complexity (understanding the processes and contexts that create the hazard in the first place as well as societal responses to it) increases as one moves from the micro (local) to macro (global) scale.

More often than not, we consider technological risks as those posed by some industry or some activity that we voluntarily engage in such as smoking or using birth-control pills. There is a direct link between the hazard source and human health implications. These are the more traditional hazards that individuals and society face. There is, in fact, a spatial manifestation of the hazard (Zeigler, Johnson and Brunn 1983).

Less direct linkages, such as the pollution caused by urbanization, resulting in significant changes to regional ecosystems which in turn affect global systems, are termed 'global hazards'. Climate change, soil degradation, and deforestation are examples of global hazards that are directly and indirectly related to the manipulation of technology. Global hazards can be distinguished from the more traditional ones because of their diffused or dispersed effects at the planetary scale—they threaten the long-term survival of the plant. In other words, their geography is more complex. They are not rare, discrete events but develop over a long period of time. Global hazards are cumulative in nature and are the end result of centuries or decades of human manipulation of technology to control nature and exploit its resources.

The processes that give rise to global hazards occur everywhere, everyday, be it in Amazonia, Bratislava, Calcutta, or your home town. Taken individually, environmental degradation (driven by technological manipulation that in turn is driven by urbanization, population growth, or poverty) has significant impacts at the regional scale. Deforestation in Amazonia is a good example of these regional impacts. Viewed collectively, these regional impacts (deforestation not only in Amazonia, but in Central America, Southeast Asia) illustrate a global pattern, creating a global hazard (contributing to global warming through the production of greenhouse emissions) that could fundamentally alter global systems and functioning. This is why scale is so crucial in understanding technological hazards.

Scale is also why our definition of technological hazard needs to be broadened beyond those rare events involving some failure in a technological system, resulting in local contamination in a relatively concentrated area. We need a broader vision of technological hazards, a vision that has as its central tenet the interaction between technology, society, and environment all of which contribute to the hazards to everyday life (Waterstone 1992).

Geography also helps us to understand the distributive impacts of environmental hazards. What is the geographic pattern of hazards? Who bears the burdens of technological hazards and where will these burdens be felt the most and why? These questions are central to any inquiry into the nature and character of technological hazards.

Old *versus* new hazards

Technological hazards are clearly related to industrial development and the use of science. As such, they are not particularly new events, but have a long history dating back to the Industrial Revolution, although others trace the origins of risk analysis and risk management to 3200 BC in the Tigris-Euphrates valley (Covello and Mumpower 1985). For example, the dangers of acute lead poisoning have been known for centuries. Symptoms suggesting lead poisoning are described in the Bible and the Talmud. Wine made and stored in lead casks has been blamed for the fall of the Roman Empire (Nriagu 1983). Charles Dickens wrote about the risks of lead poisoning and the choices workers made to accept the risk in order to earn a daily wage in his *The Uncommercial Traveller*, first published in 1860.

While the effects of acute lead poisoning or plumbism are well understood, it was not until recently (1970s) that the medical community discovered that very low levels of lead in the blood can cause problems especially in young children and developing foetuses. Inner-city children in the US were disproportionately affected from the primary sources of contamination: leaded gasoline and leaded paint. The public policy response was to reduce and phase out the lead content in gasoline and paint. However, pre-1975 housing still had lead-painted walls and ceilings. When white urban professionals began moving back to the cities and gentrifying homes during the 1980s, the hazards of lead paint were 'rediscovered'. Parents were removing the paint (mitigating one hazard) but were creating another (leaded dust) in the process. Children were once again exposed, resulting in more aggressive mitigation measures (professional remediation for leaded paint) and lead screening for children (Heaney 1992).

We have not yet discovered all the technological hazards that plague modern societies. In fact, we 'discover' new hazards every day as the lead example illustrates. We know even less about mitigating their impacts. One recent debate concerns high-voltage transmission lines and their role as possible human carcinogens. The link between low-level electromagnetic fields, birth defects, and various forms of cancers, especially leukemia and brain tumours, has been documented (Brodeur 1989; Hester 1992). Since it takes decades for cancers in particular to develop, we really have very little information on the true consequences of these hazards. The same is true for the low-level radiation emitted from colour television sets, computer terminals, and other technologies that govern our information society.

Another continuing debate is over the health impacts of low-level ionizing radiation. Recent reports from the National Academy of Science's Committee on the Biological Effects of Ionizing Radiation (BEIR V) acknowledge studies showing increased leukemia and cancer rates from low doses of nuclear fallout from above-ground weapons testing, nuclear power plant accidents, and diagnostic X-rays. The permitted levels of radioactive releases are now being challenged and revised downward as the longer-term health consequences of the nuclear age become more apparent (Gould and Goldman 1991).

The use of the herbicide 'Agent Orange' in the defoliation campaign during the Second Indochinese War provides a third example. Exposed Vietnam veterans who are now suffering from the adverse effects of dioxin contamination have filed class action suits. The social and legal significance of this instance of mass exposure to toxic substances has now changed much of the legal thinking about civil law and the ways in which victims can recover damages from chemical manufactures as a result of inadvertent exposures (Schuck 1987). It has even prompted a new specialty within law, toxic torts—a revision of common law that allows plaintiffs to claim injuries (health, property, environmental) resulting from industrial activities (see Chapter 4).

Establishing a scientific 'ethos' of risk

It was not until the end of World War II and the beginning of the atomic age that more public concern was registered about the increasing hazardousness of society. By the 1970s, this awareness was quite pronounced. Between 1957–1978, for example, the US Congress passed more than 178 laws on technological hazards and their management including the enabling legislation for the Environmental Protection Agency (EPA), Occupational Safety and Health Administration (OSHA), and the Consumer Product Safety Commission (CPSC) (Kates and Kasperson 1983; Johnson 1985).

Within academic circles, environmental hazards research began in two social science disciplines: geography and sociology. As a student of Harlan Barrows, Gilbert White practised geography as human ecology, examining the relationship between natural hazards and societal responses to them. He established centres for natural hazards research first at the University of Chicago, and later at the University of Colorado, drawing not only from the geographical community but from all the social sciences: psychology, economics, sociology. The development of a cognate field, technological hazards, was the logical outgrowth of the natural hazards research agenda, for many of White's students felt these hazards had a larger impact on society (mortality, social costs, ecosystem productivity, species extinction, pollution) than their 'natural' counterparts (Harris, Hohenemser and Kates 1978).

In 1963, sociologist E. L. Quarantelli established the Disaster Research Center (DRC) at Ohio State University, continuing his research into the sociology of disasters. Initially driven by pragmatic concerns about civil defense in wartime, the DRC expanded to include how organizations functioned during disasters in both peaceful and wartime conditions. Case studies which examined how communities responded and evaluated social institutions that governed recovery from natural disasters or technological accidents were the hallmark of the centre's work (Quarantelli and Dynes 1977; Quarantelli 1987). The DRC provided most of the training of entire generations of disaster sociologists and monopolized sociological research in the field during the 1960s and 1970s. In 1985 the Disaster Research Center found a new institutional home at the University of Delaware, where it continues its active research tradition.

A milestone in stimulating modern research on technological hazards was the 1969 publication of Starr's article on social benefit and technological risk in *Science*. This comparative analysis examined the voluntary/involuntary nature of public exposure to risk and the willingness of society to trade-off risks for larger societal benefits. It called for a more rigorous approach to making these societal choices and understanding the true extent of the technological risks. The focus was clearly on risk, not hazards.

The need for further research into technological risk and hazards was clear and Starr's publication led to a number of multidisciplinary research efforts and workshops in the National Academy of Sciences, but more importantly in the National Academy of Engineering as well. Of particular concern were issues of the role of science and scientists in determining safety (Lowrance 1976). Within the National Science Foundation a programme on Technological Assessment and Risk Analysis was established in 1977 and still exists today as a programme within the social, behavioural, and economic sciences directorate.

Conferences sponsored by the National Academy of Sciences, the National Science Foundation, and the National Academy of Engineering helped stimulate interaction among hazards researchers in many disciplines. Professional societies (American Sociological Association, Association of American Geographers, International Committee for Research on Disasters) also sponsored sessions at their annual meetings. In 1980 a new interdisciplinary society was formed, the Society for Risk Analysis (SRA), that draws membership not only from academe, but from government researchers, regulators, and private industry as well.

The bifurcation of the community into hazards researchers and risk professionals is largely derivative of methods used to understand and manage risk and hazards. Risk managers prefer quantitative risk assessment (QRA), a mathematical technique designed to forecast risks that cannot be measured directly such as the health effects of chemical exposure (see Chapter 3). Viewed by many as scientific truth, the inexact art of quantitative risk assessment is now the principal tool by which regulators make decisions regarding the acceptability of risks (Wartenberg and Chess 1992). Hazards researchers prefer a more eclectic approach employing a wide range of approaches: spatial, contextual, historical, narrative, and quantitative. Their science doesn't always produce a numeric that distinguishes high risk from low risk (everything above this value is bad, everything below is good). Rather it allows for comparisons and a fuller understanding of the causes and consequences of hazards.

Contested views

The traditional view of hazards as extreme events originating from natural phenomena or technological failures is being replaced by a broader-based view of environmental hazards. The distinction between resources management and hazards is also blurred as is the separation of hazards into natural, social or technological. Chronic pollution problems or soil degradation that aggravate the human condition over time and space are just as much a technological hazard as is the meltdown of a nuclear power plant. Similarly, a focus on more global environmental problems and the processes and responses to global change in a range of systems (social, technological, natural) provide challenges for the best environmental hazard researchers.

This book takes the view that technological (environmental) hazards are socially constructed. They are products of failures in technological devices or systems as well as failures in political, social, economic systems that govern the use of the technology. The result is a complex set of differing responses to hazards that are as variable as our perceptions of the risks we face. Individual and societal reactions to the hazards are further enhanced or constrained by social, economic, political, and scientific institutions. Since the fundamental nature of risk and hazard is often contested in the first place, it is no wonder that our responses to these hazards are equally politicized. There are also egregious inequities in the impacts of hazards and in the burdens of risk and hazards both on people and places further politicizing risk and hazards. Finally, there is a paralysis in management with these contested views of risk, hazards, and their impacts that often leads decision-makers into making what I call tragic choices.

One's own perspective becomes increasingly important in these debates over risk and hazards as it provides the prism or lens through which we view them; it also colours our actions in response to these environmental threats. We are increasingly confronted by how our own environmental philosophies (e.g. technocentrism, deep ecology, bioregionalism, ecofeminism) influence individual views and collective actions and the acceptability of public policies. Many argue, for example, that scientific knowledge is socially constructed (Harding 1991). Science is just one way of viewing the world, not the best, nor the only, just one. Thus, we can think of science and scientific knowledge as an attitude or viewpoint where social and political considerations enter into the production of scientific 'facts'. It is no wonder then that different groups of scientists working in similar circumstances often produce radically different 'facts'.

Technology can be viewed in a similar manner. As Wajcman (1991) argues, we need to view technology as a form of social knowledge, practices, and products. It is the result of conflicts and compromises, the outcomes of which are dependent on the distribution of power and resources between different groups in society. Rather than framing the question as a dualism—the problem is either man's monopoly of technology or women's technological incompetence—Wajcman suggests that women demystify the 'technical expertise' myth by gaining access to knowledge and institutions. Instead of running away from majors in science and engineering, we should be running towards them. Science and technology are not gender neutral nor are they apolitical. Thus, our fundamental vision of science and technology is already challenged. These contested views ultimately influence how risks and hazards are identified, assessed, and managed and are important to our understanding of how to live with risks and hazards.

The first half of this book provides more introductory material on the perception of risks and behavioural responses (Chapter 2), public policy formation and determining acceptable levels of risk (Chapter 3), and a review of management systems and constraints in mitigating risks and hazards (Chapter 4). The remaining chapters illustrate these themes using specific case studies. For example, Chapter 5 illustrates the importance of context (historical, economic, political, spatial) in understanding the production of hazards from the chemical industry. Chapter 6 examines the complexities of geographic scale in hazards management. Transboundary problems in managing the export of hazardous waste and hazardous technology, including ethical and equity issues are the focal point. Chapter 7 uses nuclear power hazards (civilian and military) to illustrate the dichotomy between the technical assessment of risk and the public's acceptance of it. The concluding chapter offers an overview of theoretical approaches to risk and hazards and how theory can be translated into practical strategies for risk reduction and hazards management. In sum, this book is about how you and I create hazards, assess their impacts, make choices about which ones are acceptable or not, and then how we learn to manage or live with them.

2

Scare of the week *risk perception and behaviour*

Has there ever been, one wonders, a society that produced more uncertainty more often about everyday life? It isn't much, really, in dispute—only the land we live on, the water we drink, the air we breathe, the food we eat, the energy that supports us. Chicken Little is alive and well in America (Wildavsky 1979, p. 32).

During the spring of 1989, there were a number of events (the Alar controversy, tainted Chilean grapes) that forced the American public to question the safety of its food. As part of the media coverage (some would say hype), *Time* magazine ran a cover story on food safety and risk assessment prompted by such questions as: 'Is anything safe? How two tainted grapes triggered a panic about what we eat. Do you dare to eat a peach? or an apple or a grape? The fruit panic was a lesson about terrorism—and living with risk' (*Time*, 27 March 1989). The excessive public fear of these risks and others such as irradiated food, when the scientific evidence doesn't warrant such concern (at least at the moment), or the excessive complacency of the public regarding other risks such as secondary smoke when science does warrant fear is the subject of this chapter. Why are some risks feared by the public and not others? Why, when faced with seemingly comparable risks, do individuals perceive them differently, and thus adopt varying coping strategies? What accounts for the divergent views of hazards by the public and the experts? How can we make more informed judgements regarding risks and hazards in modern societies, including the next scare of the week?

Recent public opinion polls solicited American views on the seriousness of a range of environmental problems in an effort to compare the perceived ratings to the 'scientific' assessments of the risks posed by them. Not surprisingly, Americans are becoming more environmentally aware. In 1984, for example, environmental pollution was rated as a very serious threat by 44% of the respondents; this percentage rose to 62% in 1989 (Dunlap 1991b). The majority of respondents to the federal Environmental Protection Agency sponsored polls evaluated 17 of the 29 listed environmental problems as very serious (Miller and Keller 1991). To these respondents, the most worrisome problems were hazardous-waste sites, water pollution from industry, worker exposures to toxic chemicals, oil spills, ozone-layer depletion and radioactivity from nuclear power plants. According to the same poll, environmental problems posing relatively low risks include radiation from microwaves and X-rays, and indoor air pollution.

Hazardous-waste sites ranked at the top of the list of most serious problems by the public (84% feel it poses a clear threat) (Dunlap and Scarce 1991), while experts placed the risks from hazardous-waste facilities at a much lower level (Miller and Keller 1991). Indoor air pollution including cigarette smoking is another hazard that experts worry about, yet it has garnered very little public concern.

Why are there such divergent perspectives? Perhaps the language used in the questions—say hazardous waste—imparted some bias. Media coverage might have played a role as both print and electronic media provide more vivid and longer-lasting images of environmental degradation. The public, when questioned about its perceptions, recalls these images more often than remembering scientific estimates or data that contradicts them. Also, success is rarely visible; failure clearly is as oil spills or airline crashes attest. Headlines in newspapers or lead stories on the television news all help to create the scare of the week in the public's mind. The gap between expert and lay perceptions is widening as the public becomes more aware of the risks of everyday life and are demanding action to reduce these risks.

Fig. 2.1 Risk reduction and creation. In reducing the impacts of environmental hazards, we inadvertently create other more harmful hazards. Used with the permission of Carl Hennicke.

During the 1970s environment degradation (the euphemism for environmental risks) was initially viewed as a threat to our local quality of life, but 20 years later public opinion and policy have changed. Environmental degradation and risks are now seen as global problems threatening the basic existence of the planet and all species that inhabit it, not just our own. Today the dominent view in American society (one that is rapidly gaining credibility elsewhere) is that we face more risks now than in the past, future risks will be greater than they are at present, and reducing one risk often leads to the creation of another (Fig. 2.1).

The axiom that risks and environmental degradation increase as society becomes more technologically advanced is a widely held view. Is this really true? Or are we simply more aware and thus more afraid of the risks of modern-day living? As we shall see, our individual and collective perceptions of risk not only influence our acceptance or tolerance of a technology or activity, but they ultimately effect public policies and the basic functioning of societies. The pursuit of a zero-risk society has enormous consequences for our political, social and economic institutions (Slovic 1987).

What is risk perception?

When a person says that a particular activity or technology is risky, what do they actually mean? How do we measure or gauge public opinions about these risks? What techniques do social and behavioural scientists use to measure risk perception and how successful are they?

To facilitate discussions among social scientists, the terminology most often used to describe how we think about risks and hazards is 'perception', and the one that we will use. Perception, in the narrow sense, is the actual receipt of environmental stimuli by one of our five sensory perceptors—sight, smell, hearing, taste, touch. Cognition, on the other hand, is the process of making sense of the stimuli that are coded and filtered through our individual experiences, value and belief systems, and personalities, and then ultimately stored as knowledge and memories. Cognitive processing goes on without us ever realizing it.

We should keep in mind, however, that there are subtle and not so subtle distinctions in the term 'risk perception' based on disciplinary perspectives. These often result in completely different research questions and methodologies, despite the mutual concern with risk perception.

In the geographical vernacular, for example, hazard perception is a process that links individual judgements of the degree of danger (risk) to action (O'Riordan 1986). This definition clearly articulates the view that perception is more than just a cognitive process that forms perceptions (more the psychological view). Hazard perception links judgement to action and examines those factors that influence the individual's choice of adjustments (or actions) in response to natural hazards. This distinction is important. Psychologists dwell on the process, while geographers focus more on the response. In particular, geographical interest in hazard-perception studies was used to explain the range of choice of adjustments in order to effect changes in public policy, not to understand the cognitive processes that underlie perceptions in the first place. In effect, geographers by studying hazard perception seek to understand why people take action in response to environmental threats and how they form their perceptions of the range of actions available to them.

Early roots

As mentioned in the previous chapter, some of the earliest perception studies were conducted by natural hazards researchers, primarily geographers. Heeding the call by White (1966) to incorporate public preferences and knowledge into natural-resource decisions, researchers undertook field studies to measure the public's perception about a number of natural events. Most of this early work examined attitudes and knowledge about floods (Roder 1961; Burton 1962; White 1964), tornadoes (Baumann and Sims 1972) and drought (Saarinen 1966) in specific locales. They combined standard geographical techniques of mapping and spatial analysis with the newly emerging field of survey research, through interviews of local residents and decision-makers. The infusion of social science research techniques (primarily from psychology) helped to stimulate even greater interest in perception studies during the late 1970s (see Whyte 1986 for an excellent overview).

In retrospect, the findings of some of the early hazard perception work seem rather obvious and commonplace such as personality and experience with the hazard influence the adoption of preventive measures to reduce the threat. In fact, the research attests to the robustness of the findings and how much of this scientific information has become part of the popular literature and public knowledge.

Implicit in all of the hazard work is a fundamental concern with the interaction between the physical (or technological) systems and the human-use systems. For example, the hazard-adjustment model proposed by Kates (1971) used three elements: the physical environment, human-use system, and environmental decision-makers. The model was used to identify the range of theoretical adjustments (how people could cope), actual adjustments (how people did cope) and the interaction between the two (why one coping strategy was selected over another) in response to natural hazards. In effect, the model helps explain the choice of adjustments that individuals and policy-makers select.

In a review of natural-hazard perception studies, Burton, Kates and White (1978) summarize the findings of a decade of investigations. While the theoretical orientation of the human-ecological approach has been challenged (see Chapter 8), the results of the hazard-perception field studies provide a rich source of local information and common understanding of how people feel about extreme events. For example, people do not recognize that they live in hazardous environments whether they are urban dwellers living in a floodplain or African agriculturalists living in drought-prone regions. Moreover, collective judgements on the likelihood of damage or injury resulting from a hazard event are no different than expert judgements for the more probable events. But, for the less probable events, the public and expert judgements diverge. Hazard-probability judgements by individuals ignore uncertainty altogether, reducing low probabilities to zero or passing the blame to others (God, Gaia, government, or all three). Knowledge of the probability of hazard occurrence in explaining adjustment works well with hazards that have a recurrence interval that is within memory of residents. It is less of a factor in explaining coping behaviour to hazards with longer time intervals between events (earthquakes, nuclear power plant accidents), less clearly recognizable precipitating events (air pollution, drought, global warming), or those hazards derived from technological sources (hazardous-waste contamination, pesticides) (Whyte 1986).

Awareness of prevention strategies makes little difference since most people know of a number of ways to reduce their losses. The actual adoption of prevention measures, on the other hand, is dependent on past experience with the hazard in question. If the area or the individual has experienced the particular hazard, they are more likely to adopt prevention measures such as insurance (individual level), or zoning (community level). Finally, the evaluation and ultimate selection of mitigation measures is influenced by experience, social communication, and understanding of and trust in the warning messages, emergency managers, and post-disaster relief organizations.

From hazards perception to risk perception

Hazard-perception studies surveyed residents where they lived about the events that were going on around them. They provided extremely detailed localized analyses of how people felt about hazards and what they were willing to do about them at a specific time (now for instance or after the next event). In this respect, these analyses of perception and behaviour were targeted more toward individuals and social groups who actually had to cope with the flood, drought or extreme weather event.

Risk-perception research took another path, focusing on experimental studies under controlled conditions, usually in college settings. The split between the field studies of hazard perception and the experimental ones on risk perception are grounded on methodological differences, as well as on distinctions in the types of theoretical questions posed. Hazard-perception work utilizes perception as an explanatory variable in explaining differences between the theoretical range of adjustments to hazards and the actual selection of coping responses. It also helps explain the differences between past experiences and future expectations of the hazard and its consequences. Experimental studies, largely conducted by psychologists, are more interested in cognitive processes including how information is processed (heuristics and biases) and how attitudes adjust based on conflicting information and differences between thought and action (cognitive dissonance). Because of the interest in process rather than overt action, psychological risk-perception research is rarely applied to real-world situations. Similarly, the findings of these experiments are rarely verified by a much larger and more representative sample of the population.

Risk perception, as a clearly articulated avenue for research, made its debut with the publication of Chauncey Starr's (1969) influential article on social benefit *versus* technological risk. In posing a fundamental question—how safe is safe enough?—Starr reacted against a strictly economic view or rational view of choice. He claimed that society arrives at a balance between risks and benefits by trial and error. But how are risks and benefits measured? According to Starr, the acceptability of risk is measured by comparing risk (which he defined as fatalities per person per hour of exposure) and benefits (defined as either the average amount of money spent on the activity by a person or the average contribution of an activity or technology made to a person's income) over a wide range of technologies and activities. Calling this the revealed preference approach, Starr argued that this method yields greater understanding of the processes by which society makes risk-benefit calculations about technologies and activities. It clearly had many more advantages than the traditional cost-benefit analysis methods used exclusively in many public policy decisions at the time.

Critiques and praises of this revealed-preference approach were widespread, coming mostly from the behavioural sciences rather than economics (Fischhoff *et al* 1981; Otway and Thomas 1982; Otway and von Winterfeldt 1982; Schwing and Albers 1980). It was clear that an alternative perspective on how to gauge public views of risk was needed, but the concern was about measurement. Rather than relying on deductive logic and inference, why not simply ask the public its view on the risks of modern society? Thus, the revealed-preference approach gave way to expressed preferences—the seemingly direct solicitation of public opinion that queried what level of safety was acceptable to the public for a wide range of technologies or activities. In other words, what level of risk were individuals or groups willing to bear? Furthermore, whose risk perceptions should we solicit?

Whose perceptions are we measuring?

The initial focus of hazard-perception research on the individual with subsequent generalizations to societal levels based on individual behaviour led to many criticisms of this approach. There were concerns as well about individual models of behaviour, especially the reliance on the bounded rationality model. Individual choice was viewed as a function of the subjective assessment of the utility (to use the economic construct) or the probability of expected outcomes regarding a specific action. The bounded rationality model (following Simon's satisficing models) suggested that people would choose the outcome or action that was good enough but not optimal. They would decide on what they could live with rather than trying to maximize their utility or select the action that would provide the greatest benefits. This satisficing strategy, despite its improvements on the 'economic man' (*sic*) model of rational choice, long touted by the economists, still generated critics. Behavioural perspectives were largely absent. Furthermore, the bounded rationality model did not address constraints on the range of choice imbedded in social, political or economic systems. Nor did it incorporate the role of social groups and social relations, and culture as a factor in mediating responses to hazardous situations.

Despite these critiques, hazard-perception studies spawned an area of intellectual inquiry that now spans many disciplines. Each has made significant contributions to our understanding of risk perception, although the focus (individual, group, society), scale (local, national, international), time frame (past, present, future) and fundamental research questions vary widely among them.

Individual choice: errors in risk estimation

Perhaps more than any other discipline, psychology has, over the last 20 years, advanced our understanding of cognitive structures and their role in the risk-perception process. Knowledge theory (what we know or learn about technology or risk) forms the basis for the technical or rational approach to decision-making. Personality theory (distinguishing between risk-taking and risk-adverse behaviour) provides the basis for normative or value perspectives on decision-making. Both help us to clarify the cognitive structures that govern how risks are perceived. Decision theory has been especially useful in matching decision-making processes at both the individual and organizational levels to more normative models of rational reasoning (Janis and Mann 1977). Heuristics or cognitive rules of thumb help us to understand the biases in our judgements (Kahneman *et al* 1982). All of this assumes, of course, that we can actually quantify or adequately measure this phenomena called risk. There is some debate over this.

In a review of the technological-risk literature, Covello (1983) highlights three primary findings relating to bias, estimation, and believability of risks. First, there are limits to human intellects that force people to apply a series of decision-making rules (heuristics) as a way to handle complex information. Often this results in the over-simplification of risks. For example, we tend to judge an event as more frequent if instances of it are easy to recall. The scattered debris of airplane fuselages, twisted metal, human bodies and rescue workers searching for the infamous black-box flight recording device are well-known images of airline disasters, an event perceived as relatively frequent. Yet, the actual occurrence of airplane accidents is relatively rare. In the United States, the risk of dying from an airplane crash is actually less than the risk of dying from an automobile crash.

In addition to information biases, we assume that similar activities and events, such as nuclear power and nuclear war, have the same characteristics and therefore the same levels of risk. In each of these examples the availability of information (easily recalled or highly visual images of events) and representativeness (similar activities equal similar risks) result in biased risk estimates. This often leads to erroneous conclusions that aircraft accidents are more frequent than they actually are or that the risks of a nuclear power plant accident are the same as a full-scale nuclear war (they are not). As a result people have been shown to be poor probability estimators, consistently over-estimating the frequency of rare (low probability/high consequence) events like nuclear power plant accidents and under-estimating the frequency of common (high probability/low consequence) events like lightning strikes.

The second major finding of the risk-perception literature is most people are generally over-confident about their risk estimates and this often leads to errors in risk estimation (Covello 1983). More important, however, this over-confidence leads to complacency, or the view that it can't happen to me. It means that we tend to under-estimate those risks that are familiar to us and under our immediate control such as driving an automobile or smoking cigarettes. Conversely, we tend to over-estimate those risks that are new to us where we feel little is known about them and ones that we are involuntarily exposed to (contaminated water or air pollution from incinerators).

Third, the psychology literature has helped to understand the differences between the experts' estimate of risk *versus* those of the public. Experts often use quantitative assessments based on the probability of the occurrence multiplied by the consequences. Not surprising, expert assessments are highly correlated with expected annual fatality rates from the activity or technology in question. Lay perceptions routinely involve more qualitative assessments and are more expansive in including factors other than fatality rates into the judgements.

Psychometrics

If we assume, for the sake of argument, that we can adequately define risk and all agree on the concept, then how do we go about measuring people's perceptions of it? The psychometric paradigm developed by Slovic and colleagues in the late 1970s (Fischhoff *et al* 1978; Lichtenstein *et al* 1978; Slovic *et al* 1980; Sovic *et al* 1985) is one widely used method. This psychometric paradigm utilizes a taxonomy of hazards to produce quantitative measures of risk perceptions and attitudes or, more precisely, a cognitive map of risk (Slovic 1987). These studies concluded that risk perceptions are in fact quantifiable and somewhat predictable.

The psychometric paradigm claims that risks are viewed as more acceptable if they are familiar, controllable, have a low catastrophic potential and/or are equally shared. Two-dimensional 'maps' illustrate commonalities in the attributes of a series of risky technologies and activities by plotting dreaded risks on one axis and unknown risks on the other. Those risks that are unknown and dreaded such as nuclear reactor accidents and DNA technology are found in the upper right quadrant (Fig. 2.2). Similarly, those risks least feared like smoking appear in the lower left quadrant.

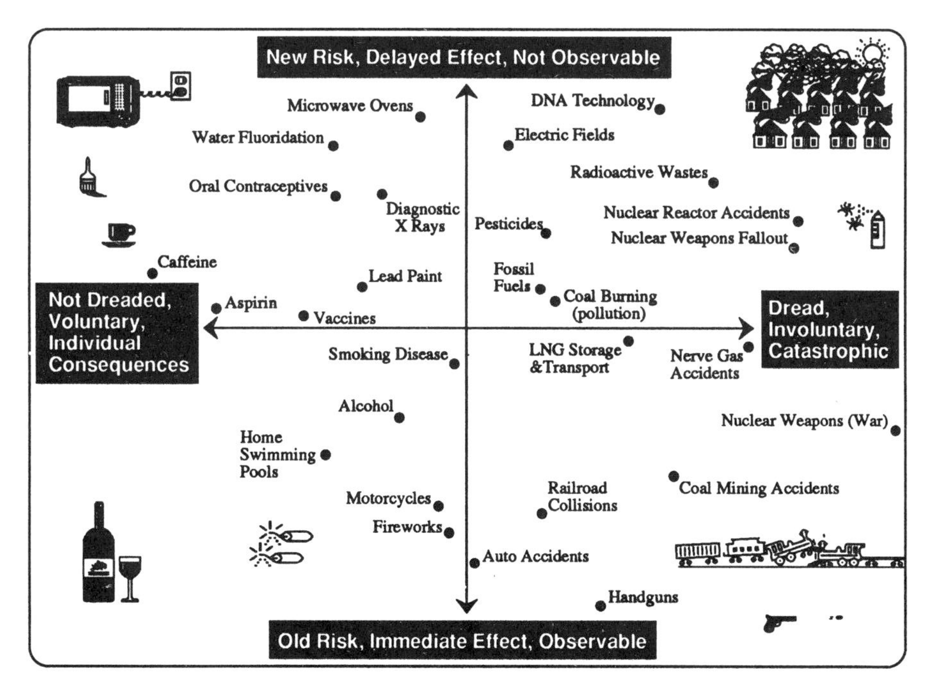

Fig. 2.2 Two-dimensional risk characteristic map. Risks that are most feared and dreaded are new, not observable, have delayed effects, and have globally catastrophic consequences. These risks are located in the upper right quadrant on the map.

Another recent development is the concept of a signal potential or the social impact or social disturbance caused by a particular activity or technology (Slovic 1987). Based on the characteristics of the hazard and its location in the two-dimensional factor space, some technologies or activities produce a higher signal value than others. For example, a train wreck may result in the loss of individual lives, but the accident occurred in a familiar and well-known system and the public took little notice. In contrast, a small accident in an unfamiliar system such as the chemical industry may cause greater social disruption and angst, despite fewer fatalities or injuries. We are unfamiliar with the chemical industry, so incidents from this system are often seen as a harbinger of even greater mishaps, including perhaps a catastrophe. The signal potential from accidents directly relates to its social impact.

The experimental focus in psychology has led to many critiques of the psychometric paradigm in particular and a reappraisal of the risk-perception research in general (Renn and Swaton 1984; Gardner and Gould 1989; Hansson 1989). The limitations cover five broad areas. The first involves the representativeness of the samples used in the psychometric paradigm research. Largely derived from empirical studies of small, non-randomly selected samples of college students in Oregon and elsewhere (or the League of Women Voters in Eugene, Oregon), there is concern that the data are not applicable to, nor do they represent, broader spectrums of the public. Despite the stated population sampling caveats in the original studies by Fischhoff, Slovic and colleagues, subsequent researchers simply took the findings at face value. As a result, the entire risk-perception literature has managed to evolve with very little replication and challenge of these basic studies using more representative sample populations stratified by race, gender, income, age or educational levels.

Another second criticism is that the psychometric paradigm entails a comparison of risky technologies and activities among one another based on the aggregated or average rankings of all individuals (Gould *et al* 1988). In this regard we are not viewing one individual's perception of risk, but rather an average ranking of riskiness of the 33 or so technologies or activities compared to one another. By aggregating individual-level data to compute variables like the mean perception of risk, or attributes of risk, we are unable to relate the sociodemographic characteristics of each respondent or measures of individual attitudes/values to their perceptions of risk.

Third, the original studies were not issue-specific; preferring to target a wide range of technologies and activities for comparative assessments of riskiness. This criticism has been muted recently, as psychometric analyses of perception of specific risks is burgeoning in such areas as nuclear power (McDaniels 1988), radioactive waste repositories (Kunreuther, Desvouges and Slovic 1988), electric and magnetic fields (Morgan *et al* 1988), and health issues (Hardin and Eiser 1984).

Fourth, the basic definition of risk is problematic. Risk, as used in these studies, is the likelihood of dying from the technology or activity. There are no other aspects of risk that are considered, such as long-term injury but not death, environmental consequences of the activity and so on. Despite its initial interest in the acceptability of risk question, the psychometric paradigm provides only a cursory view of this aspect. Instead it emphasizes relative risk rankings and risk attributes within narrowly framed lists of technologies, activities, and descriptive characteristics.

Finally, the psychometric paradigm has failed to relate risk perception to individual action or social policy formulation (Kishchuk 1987). The link between individual risk perception and willingness to do something to reduce risk from the technology or activity is not addressed by this technique.

The social fabric of risk

Who participates in creating, filtering, and allocating risks? How do these processes occur? The transformation of risk away from probability assessments to more fundamental concerns on how risk questions are framed and the acceptability of the answers has been enlightened by social science inquiry during the last decade.

Sociology has helped focus attention on both the individual as well as groups and the social mechanisms that influence risk perceptions. Social psychology, for example, has contributed to our understanding of social values, norms, social roles, and their connection to personal judgements. On a larger scale, sociological inquiry has helped to explain how groups or communities respond to risk and the role of social values, institutional constraints, referent group judgements, communication, and power relations in enhancing or constraining social actions. Social influences (the role of family friends, workers, and some public officials) have been shown to strongly influence actions, often providing after the fact rationales for hazard behaviour (Short 1984). The spontaneous evacuations during the Three Mile Island accident provide good examples of the role of social influences in mediating behavioural responses. One of the primary reasons for evacuating for many residents was that their neighbours were doing it and they did not want to be the only household on the block to remain. This behaviour was only partially related to their individual assessment of the risks (Cutter and Barnes 1982).

Sociologists have asked crucial questions on the salience of risk. How important are risk perceptions or judgements when compared to other social problems like crime or unemployment? This is not to trivialize risk perception, rather it aims to place the public's assessment of risk in context with a host of competing social issues and concerns. Certainly, people may feel a heightened sense of risk, but compared to the other social problems they confront on a daily basis, technological risk has a lower priority than unemployment or homelessness, for example.

Broader social science inquiries have stimulated more field verification of psychological data, expanding and enlarging sample populations. Rather than relying completely on statistical measures of risk, sociology has helped to frame risk-perception research in a broader context and refocus it away from probabilistic functions or psychometric analyses. How people live and how day-to-day activities influence perception and behaviour is one of the main contributions of the social sciences. In fact, sociologists, more than anyone else, have helped to draw attention to risk as a social construct where 'after the fact judgements' are often used as a mechanism for rectifying public action and public attitudes (Dietz, Frey and Rosa 1991). The duality of the social context of risk and the social fabric at risk (Short 1984) have helped stimulate a new understanding of how individuals make risky decisions and how society responds to risk. According to Short (1984, p.719) living with risk is more socially than technically driven.

Perhaps, the greatest achievements in sociology have been at the organizational level: how social systems function and how decisions are made or not made. Contributions such as Perrow's (1984) on the character and nature of complex technologies, how risky systems all fit together and interact with human systems to produce hazards help us to understanding that risk does not occur in isolation. Stalling's (1990) work on the media as a social institution and its influence on how risks are selected and defined by the various media stakeholders is another area. Palm's (1990) analysis of the managerial élite (planners, emergency management officials, policy analysts) or 'gatekeepers' as she calls them, and their role in constraining or enabling hazard perception and behaviour is another example.

Cultural biases

The pioneering work of Douglas and Wildavsky (1982) championed the view of a culturally determined perspective of how societies select and deal with risks. They argued that sociocultural processes were working to govern the selection of risks (why some are emphasized and others ignored) in peasant, agrarian or industrialized societies. The cultural selection of risk was not linked to objective risk measurements or the physical reality of the risk. Rather the selection of risk reflected moral, political, economic, and power positions that were all value-laden and culturally constructed. These social relationships resulted in cultural forces that downplayed some risks while heightening others, all of which were used as cultural mechanisms for controlling the social groups. What is determined as dangerous is framed as a social problem rather than a technical one. In other words, Douglas and Wildavsky articulated the position of risk as a socially constructed phenomena.

In a slight variation on that theme, Rayner and Cantor (1987) suggest that organizational cultures influence risk perception and the way organizations manage risks. These corporate structures are a function of the power, autonomy and corporate mind-set of the organization itself, be it a private firm, government agency or special-interest group. Rayner and Cantor suggest that risk managers ultimately end up arguing over conflicts in trust and social equity in risk decisions rather than on the quantitative measures of risk. Given different cultural biases (corporate or institutional culture), views of social justice, and economic interests, it is no surprise that each organization will ultimately choose different paths in managing risks, despite similar perceptions of the risk at the outset.

In a review of risk-perception theories Wildavsky and Dake (1990) summarize five competing views that are frequently used to explain how society perceives of and selects risks. Knowledge and personality theories were discussed earlier in this chapter. Economic theory posits that the rich take risks because they benefit more from the technology or activity and are partially shielded from any adverse consequences. Alternatively, they can simply buy their way out, or are willing to pay for less risky goods and services. Willingness to pay to reduce risks is part of this economic view. For example, Hammitt (1990) found that consumers were willing to pay up to 50% more for organically grown produce in order to reduce ingestion-related risks of pesticide exposures.

Political theories suggest that risk perception is conflict between various special interests. Simply put, risk perception is a gigantic clash of conflict of interests between different ages, classes, ideologies, political parties, races, and genders. Finally, the cultural theorists explain that society chooses risks based on deeply held values, world views, and social relations. In an empirical test of which theory provided the best explanation of societal risk perception, Wildavsky and Dake found cultural theories were the best in reinforcing the findings of the earlier cultural-selection-of-risks hypothesis.

Is this all we know about risk perception?

The psychometric paradigm is the most widely used method for measuring risk perceptions, but it certainly is not the only one. As we have seen, sociologists, anthropologists, and geographers all contributed to the risk-perception literature. In addition, social and environmental psychologists working from attitude theory and change (Verplanken 1989; Gould *et al* 1988), affective responses and judgements (Johnson and Tversky 1983, 1984) and theories of reasoned action (Ajzen and Fishbein 1980; Young and Kent 1985) have helped our understanding of risk perceptions. Table 2.1 presents the major findings of the empirical risk-perception literature in simplified form. What is clear from looking at this table is not how much we know, but rather how little we know about how individuals and society perceive risk.

Table 2.1 Major findings in individual risk-perception research.

1) People simplify.
2) Once people's minds are made up, it is difficult to change them.
3) People remember what they see.
4) People can't detect omissions in the risk information they receive.
5) People disagree more about what risk is than about its magnitude (how large a risk is it?).
6) People have difficulty in detecting inconsistencies in disputes about risk.
7) People find it hard to evaluate expertise.

Source: Fischhoff 1985.

Factors that influence risk perception

There are many factors that influence how individuals, groups, and society view risks and render judgements on the acceptability of risk. We have already alluded to some of these factors such as disciplinary biases in research design and measurement, individual personality, social influences, and cultural forces. There are other factors that also deserve a bit of our time.

Experience

If a flood or nuclear power plant accident happened once in my community, will it happen again? Will my neighbours, friends, and family view the risk differently because we now know what to expect? There is a wealth of evidence that prior experience with a technology or natural hazard influences one's perception and alters responses to the events in question. In some parts of the world, natural hazards are so pervasive and the populations so experienced that a 'hazards culture' has evolved (Burton, Kates and White 1978). Risks are downplayed and coping responses are viewed as normal, just a regular part of daily activities.

The role of prior experience with a particular technology or activity clearly enhances the probability of adoption of mitigation measures or some other form of adjustment. In a number of studies on energy-conservation strategies, for example, Macey and Brown (1983) clearly demonstrated that prior experience with any form of energy-conservation activity such as lowering thermostats or putting on a sweater greatly enhanced the likelihood that the person would adopt more aggressive mechanical or institutional conservation measures such as weather-stripping, replacement of an inefficient energy provider, insulation and so on. Similarly, societal experience has led to better preparedness in anticipating natural disasters and some technological failures. The role of seat-belts in saving lives and reducing serious injuries from motor vehicle accidents has weighed on our collective experience, so much so that now most of us automatically buckle-up for safety when we enter the car.

In a study of perceptions of nuclear accident risks, Lindell and Perry (1990) compared the perceptions of residents in Washington state five months before the 1986 Chernobyl accident and one month after. They conclude that a major accident anywhere can actually decrease the perception of the threat at the local level. Respondents actually felt that accidents in their local area were less likely to occur because of the Chernobyl accident.

Lack of experience tends to amplify the risks until such time as risks are moderated or people have adapted to them. In Edelstein's (1988) study of Jackson, New Jersey, a community whose groundwater (the source of drinking water) was contaminated by leachate from a closed landfill site is a case in point. Residents at first had little experience with toxic exposures and when the contamination was discovered, they became outraged, never suspecting that by simply drinking the water, their children were at risk. The psychological stress resulted in a number of social problems (increased divorce and drinking) and many families became dysfunctional. Residents coped for years, buying bottled water in the local store or having it delivered until another potable source of drinking water was found. While many residents are still angry, they still live in the community and are a little more sensitized to toxic-contamination issues.

Cross-cultural comparisons

In the natural hazards arena, cross-cultural analyses on attitudes and adjustments have a long and distinguished history (White 1974). These cross-cultural analyses used standard questionnaires (modified slightly for the local setting) and were administered in many different countries. Their purpose was to elicit hazard perceptions and to account for variations in attitudes and choice of adjustments based on contextual factors such as social and economic status, personality variables, or local (site and situation) differences. These comparative assessments found that four factors were useful in describing the variations in attitudes and coping responses: experience with the hazard, material wealth, personality, and the role of the individual in the social group (Burton, Kates and White 1978).

There has been some work on cross-cultural risk comparisons in the technological hazards arena as well (Vlek and Stallen 1981; Englander *et al* 1986; Teigen *et al* 1988; Bastide *et al* 1989; Keown 1989; Kleinhesselink and Rosa 1991). These studies attempt to confirm that culture, in fact, makes a difference in risk perceptions. Most of these comparisons utilized the psychometric paradigm and found few differences between the perceived characteristics of risk and risk rankings, whether from Oregon students or ones in Hong Kong, Japan, Hungary or Norway. Whether or not culture has any influence whatsoever is still unclear given the shortcomings in the psychometric paradigm described earlier in the chapter. But at least the issue is being raised and a number of researchers are now employing alternative modes of risk-perception measurements in cross-cultural comparisons rather than relying solely on the psychometric paradigm (Renn and Swaton 1984).

Environmental philosophy

Environmental philosophies and ideologies help to shape individual and societal views toward nature and technology. If these underlying philosophies are placed on a continuum, the most pervasive philosophy in industrialized regions stretches from ecocentrism at one pole to technocentrism at the other (O'Riordan 1981). Ecocentrism views the world according to natural laws. It preaches responsiblity and care, with humans just a part of the much larger Gaian system. The Earth's functioning involves a complex balancing act between all parts of nature (living and non-living elements), each of which seeks stability through the ecological principles of diversity and homeostasis. The result is a view of society that small is beautiful, the title of E. F. Schumacher's famous book. Low-impact technology turns out to be the rule rather than the exception. The rejection of scientism and its implied rationality and a movement away from technological optimism are elements in this ecocentric world view and thus govern how risks are perceived and evaluated (Piller 1991; Shrader-Frechette 1991).

In polar opposition is technocentrism, the dominant view since the Industrial Revolution. Rationality, science and managerial mastery form the underlying tenets of this philosophy. Nature is there to be used for the betterment of society. The duality of nature and human society is established, thus producing the anthropocentric view that is so prevalent today. In other words, nature became threatening rather than nurturing and passive. Technology was used to conquer nature, to subdue it. Many of the technologies that liberated men and empowered them over nature were also used to dominate women in patriarchal societies.

Another environmental philosophy that is gaining acceptance is ecofeminism. As Cutter *et al* (1992) suggest in their discussion of feminist perspectives on technological risks, men as decision-makers traditionally undervalued the nurturing philosophy inherent in ecocentric views, feeling more comfortable with rational, scientific views not subject to emotions. Since technology is the creation of science and engineering (fields where women traditionally lacked access and acceptance), other views are rarely entertained by the professionalized élites. Women, if involved in decision-making at all, were either citizen activists and therefore dismissed as 'bunny lovers', or masked their emotional involvement under the veil of scientific expertise. Thus, risk perceptions based on anti-technocratic, anti-masculine, or anti-progress modes of thought were and still are summarily dismissed as irrational. It didn't matter if the alternative opinions were from women, minorities or the poor—they all bucked the scientific establishment and managerial élite and were consequently devalued and ultimately ignored.

Race, gender and socioeconomic status

People's background (race, gender, socioeconomic status) may help to explain variations in risk perception. We know, for example, that women, minorities, and the poor often lack political power and are consequently exposed to more technological risks ranging from the siting of hazardous facilities to birth-control technologies. There has been a pattern of environmental racism in the US. The range of environmental hazards disproportionately affecting minorities includes toxic-waste sites (Commission for Social Justice 1987; Bullard 1990), industrial pollution (Berry *et al* 1977), and pesticide exposures of migrant farmworkers (Wright 1990). Issues of social equity and who bears the burdens are codified in Bullard's (1990) analysis of the relationship between race and hazardous-waste sites. National recognition of uneven risk burdens has led to accusations of environmental racism and attacks on the US Environmental Protection Agency for its lack of environmental enforcement in low-income and minority communities (Schneider 1991a). However, what is missing from these cries for environmental justice is an examination of racial differences in perceptions and how these affect levels of environmental activism (Mohai 1990; Bryant and Mohai 1992).

In his review on the level of environmental concern by African Americans, Mohai (1990) found conflicting evidence in the literature. On the one hand, lower levels of environmental concern and involvement were found among blacks than whites. Conversely, a number of studies found that environmental concern by black residents was higher than for whites. In an early study of the relationship between community attitudes toward pollution and abatement strategies to reduce pollution levels in Chicago, it was found (Cutter 1978) that risk burdens were equally shared when more than one pollution indicator was used. Furthermore, black respondents were more likely than whites to engage in pollution-reduction measures like recycling, especially black residents in the middle-income categories. In a more detailed analysis (Cutter 1981) higher levels of concern were found among predominantly black communities than in white ones as well as more willingness to undertake actions to do something about the pollution. Mohai (1990) also found blacks to be as concerned or more so than whites about environmental quality.

One explanation for this divergence may be related to the salience of the issue, e.g. some aspects of environmental quality (toxic contamination) are more salient to blacks than others (e.g. pristine wilderness environments).

This might account for increased levels of environmental activism at the local level, especially in response to toxic-substances issues. Since blacks are often disproportionately exposed to toxic-air emissions from industry or hazardous-waste facilities, this comes as no surprise. Perhaps we have entered a new age, where environmentalism is no longer the sole purview of the white upper and middle classes in American society or elsewhere. Certainly, these studies have illustrated that race is a factor in the perception of risks.

There is also considerable evidence for gender differences in concern for environmental quality with women being more environmentally oriented than men (Milbrath 1984; Schahn and Holzer 1990). General opinion surveys also find women in the vanguard, with higher support levels for environmental regulations and policies than men. There is a similar pattern in the risk-perception literature (McStay and Dunlap 1983; Stallen and Thomas 1988; Fischer *et al* 1991). Gender, as opposed to biological sex, is a term that describes the extent to which personality traits are socially defined as masculine or feminine. According to Merchant (1981), the male psyche is based on separation, distinction, dualism. The female psyche is grounded in empathy, wholeness and identification. This social construction of identity leads to different views of nature. Thus, if women perceive and experience the world differently than men, it makes some intuitive sense they would also perceive risks differently. Likewise, if men listen to their more feminine traits, their views would reflect gender distinctions as well.

A number of empirical analyses have demonstrated gender differences in the perception of technological hazards, especially nuclear war (Boulding 1984; Silverman and Kumka 1987; Levin *et al* 1988), acid rain (Arcury *et al* 1987; Steger and Witt 1990), and industrial hazards and health (Stallen and Thomas 1988; Bastide *et al* 1989). In other studies the findings were less supportive (Cutter *et al* 1992) because of the inability to differentiate gender characteristics other than using the sex of the respondent. Needless to say, the role of gender in the perception of risks and their ultimate acceptance is an important one.

We have sketchy data on other social descriptors of risk perceptions. In most cases, the magnitude of the correlations between race, ethnicity, gender, socioeconomic status and risk perceptions are low. Thus, while we may think these are important correlates of risk perception, we still have no definitive research that allows us to make broad sweeping statements about their overall importance.

Distance

Distance has a direct bearing on the objective risks that populations are exposed to during and/or immediately following a hazardous event. Distance not only serves as an indicator of danger (closer to the plant, the greater the risks) but cognized or estimated distance serves as a heuristic anchor for judging risks and individual vulnerability to the threat (Maderthaner *et al* 1978; Lindell and Earle 1983; Shippee *et al* 1980). Cognized distance from a nuclear power plant, for example, enables individuals to personalize the risks. It provides an individual with a criterion to judge a series of stimuli such as news reports, advisories, and the behaviour of others in order to develop their own set of coping strategies (Cutter 1984). It also provides a criterion for rationalizing the behaviour that was selected.

Perception–behaviour linkages

One wonders why there is all the emphasis on perception if we can't ultimately relate our perceptions to actions of some sort. One of the greatest failings of the risk-perception literature is the unwillingness or inability to take the next step and relate our perceptions of risk to our individual and collective responses to it. We have very limited knowledge on how perceptions influence behaviour and how our behaviour influences our perceptions of risk. There have been relatively few analyses that specifically link the perception of risk to some overt action. On the other hand, the political science literature is filled with studies that relate activism to socioeconomic characteristics and the social organization of the community.

One of the problems in the social psychological research is that all too often, there is no correlation between perceptions and behaviour, or else the correlations are so low as to offer very little in the way of supportive evidence on the linkages. In a survey of residents in Connecticut and Arizona, for example, Gardner *et al* (1989) found very few correlations between activism and attitudes even when attitudes were disaggregated to attitudes toward risk in general, technology or the environment. Researchers using Fishbein and Ajzen's theory of reasoned action (Ajzen and Fishbein 1980) have obtained slightly better results (Young and Kent 1985; Lotstein 1990).

At the societal level, according to Burton *et al* (1978), human responses to natural hazards are governed by four factors: (1) the characteristics of the extreme events (frequency, duration, areal extent, speed of onset, spatial dispersion or temporal spacing between events); (2) experience creating a state of awareness of the hazard that diminishes over time; (3) resource-use based on the intensity of use; and (4) level of material wealth. The relative importance of each of these in the selection of adjustments or coping strategies varies from event to event, as well as from place to place. We can infer that responses to technological events may include this mix of factors as well as others.

This is not to suggest that the social psychological literature completely ignores attitudes and behaviour. In fact, a number of very interesting studies have been conducted on specific failures such as Three Mile Island and the behaviour of residents (usually evacuation) as it related to their perception of the threat.

Politicization of risk

Technological risk is a highly contentious and politicized issue. O'Riordan (1986) offers some general observations on the reasons.

1) Risk and culture intertwine. . . . Risk therefore has a socially moral element: exposure to man(*sic*)-caused danger is a feature of societal failure.
2) Fact and value interconnect. Because no science is value-free, risk provides a convenient vocabulary for scientific disputes and a context for intensifying scientific conflicts.
3) Institutions for fusing scientific evaluation with political judgements are proving inadequate. Risk therefore provides an avenue for frustration over secrecy and lack of accountability. . .
4) Internal and external modes of risk analysis do not always connect. How people judge risks is not always distinguishable from their political beliefs or their attitudes to authority, expertise, party loyalty or national chauvinism (pp. 292–293).

One of the more interesting questions in the risk-perception arena is the dichotomy between expert and lay (non-expert) judgements of risk and their role in public policy. There are many competing views. One widely held notion is that the public should be totally excluded from any decision-making because they are not informed enough to make rational choices (Lewis 1990). A more paternalistic view suggests that the public could provide 'informed' decisions if they were given more and better information, presumably by the experts. The latter view underscores the premise behind the entire field of risk communication and why regulatory agencies are so quick to adopt risk communication in handling public disputes over risk (see Chapter 3). A third alternative, not widely held at the moment but gaining credibility, is to believe the public's perception of risks. The reasoning here is quite simple. Since the risk-perception literature has demonstrated that all judgements, including those of the expert, are subject to bias, who is to say which set of information is better? Risk perception becomes a socially constructed concept where the public view and acceptance is as important as expert testimony. In many instances, it is actually better since it shows what the public is willing to tolerate in the way of risks. The public has a vested interest in assessing, evaluating and managing technological risks. Similarly, policy-makers should incorporate these views in their policy and regulatory strategies as well, further politicizing risk decision-making.

The 'Alar' (or daminozide) controversy is a good example of how risks are politicized. Daminozide is a plant-growth regulator commonly used on peanuts, apples, and other fruit. The apple industry in particular has been a heavy user of 'Alar', the trade name. The chemical delays ripening (thus delaying harvest time), reduces the pre-harvest dropping of apples from the trees, extends shelf life once the apples are picked, and most importantly produces redder and firmer apples (Rosen 1990). Early in 1977 the USEPA targeted Alar for review as a suspected carcinogen. By 1984, despite a number of scientific studies, the EPA had identified major gaps in the literature regarding residues, chronic toxicity, and mutagenicity of daminozide (Jasanoff 1987). In January 1986 the EPA announced the permitted continuance of daminozide, pending the completion of further toxicological and residue analyses by both the EPA and the manufacturer, Uniroyal. The risk-assessment data at the time were inconclusive and the agency simply wanted more data before pulling the chemical off the market.

What politicized the issue for the public (a scientific controversy was already occurring) was the 1989 release of a Natural Resources Defense Council report on the disproportional effects of daminozide on children. Children are the primary consumers of apples and based on their body weights and greater than normal consumption of apples in the form of juices and applesauce, the NRDC report claimed that children were consuming large quantities of the chemical. Most of the toxicological studies, in fact, were based on adult apple-consumption patterns and body weights. CBS television's '60 Minutes' programme ran an exposé on 'Alar', branding it a potent carcinogen and a major threat to children's health. Throughout the programme, the skull and crossbones symbol for death was prominently displayed. Meryl Streep, the actress and concerned mother, was enlisted to help heighten public awareness, testifying before Congress on the dangers of 'Alar'-treated apples to the nation's children. The media campaign (Fig. 2.3) that ensued was instrumental in heightening public awareness and ultimately in forcing EPA to ban daminozide, effective immediately. Despite the evidence of 'experts', the risks were never fully elucidated, and a seemingly beneficial chemical was withdrawn from the market. Apple growers lost millions of dollars in the process. Many parents were afraid to purchase red apples or any apple products.

Apple growers filed a US $200 million lawsuit against CBS and NRDC charging that they exaggerated the risks, resulting in devastating income losses for the growers. Congress is preparing a variety of bills that would overturn the Delaney amendment (see Chapter 4) that forbids chemical residues on processed food if they are judged to be carcinogenic, even if the health risks are negligible. In October 1991 the EPA concluded

Fig. 2.3 The politicization of risk. The 1989 Alar controversy focused attention on divergent perceptions of risk and the preoccupation of Americans with a zero-risk society. Reprinted with special permission of North America Syndicate, Inc.

yet another series of toxicological tests on Alar and found that the potency level is half of the original estimate. As the controversy makes its way through the courts and Congress, Alar continues to be used worldwide. A United Nations advisory committee and a peer-review panel of the World Health Organization and Food and Agricultural Organization concluded that the small quantities of Alar residues on food posed no risk to human health. They subsequently set acceptable levels of intake (Marshall 1991).

The dichotomy between the technical assessment of risk (suspected carcinogen but needing more evidence to substantiate the case) *versus* the public reaction (hyped by the media, movie stars, and preoccupation with a zero-risk society) illustrates the politicization of risk and the need for managers to incorporate public views in risk-management decisions. Challenging the prevailing view of 'experts' by questioning the fundamental attitude embodied in scientific explanations is becoming more acceptable. Science or more precisely the scientific method is an attitude, a way of looking at the world. It has just as many assumptions as other methodologies and therefore cannot be judged as correct simply because it is science. Risk decisions, then, become matters of public discourse and debate and not solely determined by scientific exchanges and quantitative estimates of the probability of cancer from the activity or technology. Lest we forget, risk is a social construct. Given this view, how do societies determine which risks to manage and how safe is safe enough?

3

Tragic choices

> 'In primitive [*sic*] societies these choices (dollars *vs* lives) were often made by the tribal witch doctor. When the need to choose between cherished but conflicting values threatened to disrupt society, the simplest path was decision by a shaman, or wizard, who claimed special and miraculous insight. In our time shamans carry the title doctor instead of wizard, and wear lab coats and black robes instead of religious garb' (Bazelon 1979, p. 277).

There is no such thing as a risk-free or hazard-free environment despite American preoccupation with a zero-risk society. Scientific progress, beginning with the Enlightenment, created new hazards, as well as uncovering previously unknown ones. Societies have always made explicit or implicit choices on risks and hazards; between a few lives lost *versus* the overall betterment of society. Warfare is perhaps the best example of a tragic choice. As Bazelon (1979) suggests, tragic choices have long been part of human existence and are found in most societies. The challenge facing modern societies is to accept the reality that risks and hazards are part of our daily lives. We can then focus our attention away from debating the mere existence of risks to more important considerations on how much risk we face and from what sources. The questions of who shall decide on which risks and hazards we should accept, and what level of safety is safe enough should dictate public discussion and policy.

The process of societal risk and hazard decision-making is a complex one. As we saw in Chapter 2 with the Alar example, there are major differences between the public's view of risks and those of experts. Similarly, there are conflicting choices to be made in the formation of public policies to reduce or manage these risks. This chapter examines the social acceptability of risk as embodied in public policy, the balancing act that regulators continually perform, and the resulting tragic choices that must be made. These choices may minimize human injuries or deaths, but often come at the expense of other species or the environment. The balancing act to prevent environmental deterioration and protect human health and welfare is one of the greatest challenges facing the modern world. There is an extensive literature on the topic of acceptable risk, as you might well imagine. Rather than trying to review all of it, I have selected a number of themes that consistently appear in public policy discussions on the acceptance or what I prefer to call the social tolerance of risk.

How much is a life worth?

In order to even begin a larger discussion on hazards, risks, and safety, we need some basic information on the hazard or risk insofar as it is known. For example, what are the benefits of the technology or activity in question, both in the short term and long term? What are the impacts (human fatalities, human injuries, environmental degradation) of such events and how do these vary in time and space? What are the costs (economic, social, political)? How certain are we about the benefits and risks associated with the technology or activity?

A recent National Research Council report (1989) highlighted five categories of information that are needed before decisions on risk can be made. First, the risks of an option must be clearly articulated. Second, the benefits must also be stated (Table 3.1). Third, alternative options and their risks and benefits must be presented including the effectiveness of each alternative as well as the costs associated with it. Fourth, the level of certainty (or more likely uncertainty) associated with the risk and benefit determinations is required. Finally, how will these technical risk assessments be conveyed to decision-makers, in what context, and how will they be used to form policy?

The determination of societal trade-offs in policy choices has evolved over time. It has been greatly enhanced by the quantitative analysis of risks, benefits, impacts, and costs, so much so it seems, that risk assessment has become more of a science rather than an art (Wartenberg and Chess 1992). There are a number of analytical techniques that provide inputs for these policy decisions. We will review four of the most prominent: cost-benefit analysis, revealed preferences, expressed preferences and risk assessments.

Cost-benefit analysis

Cost-benefit analysis has a long tradition in environmental decision-making. In this approach, the expected benefits from a proposed technology or activity are simply weighed against the expected costs. In order to make the calculation, all the adverse consequences must be enumerated, probabilities assigned to the likely occurrence of each adverse condition, and a determination of the costs (or losses to society) should the consequence occur. A summing of these costs is then compared to the benefits that will accrue (which have, incidentally, also been assigned dollar amounts).

Aside from the obvious reliance on economic indicators and the assumption that all things have a price attached to them (such as the value of a view), cost-benefit analysis has other limitations. First, it assumes that data are readily available to quantify all the adverse consequences that are imaginable. Second, it also assumes that we can visualize all the

Table 3.1 Informational needs for risk/hazards decisions.

Nature of the risk
1) What hazards are we concerned about?
2) What is the probability of human exposure? Is it a single event resulting in acute exposures, or cumulative events creating chronic exposures?
3) What is the probability of harm or the potency of the exposure?
4) What is the distribution of the exposure and who is most affected?
5) What are the sensitivities of different populations, such as children, the elderly, women, to potential exposures?
6) What is the interactional nature of the exposure? Will it act as a blocking agent, or react with other elements to create synergistic effects?
7) What are the characteristics of the hazard itself (e.g. magnitude, duration, potential to cause death, frequency, speed of onset, catastrophic potential)?
8) What is the population at risk?

Nature of the benefits
1) What are the benefits, who benefits and how?
2) How many people benefit and for how long?
3) What is the probability that the benefits will be realized? What events might intervene to reduce those benefits and how likely are these events to happen?
4) Is there a disproportional distribution of benefits?
5) What are the specific qualities of the benefits (e.g. physical comfort, environmental quality, improved health, improved welfare)?
6) What is the sum total of societal benefit?

Alternatives
1) What are the alternatives?
2) What are the risks associated with these? Benefits?
3) How effective are each of the alternatives in reducing risks?
4) What are the costs of each alternative?
5) Are the risks, benefits, and costs evenly distributed among relevant populations?

Uncertainty
1) What are the weaknesses in the available data used to determine risks and benefits?
2) What are the assumptions and models that the data are derived from? How much dispute is there over these assumptions and models?
3) How sensitive are the estimates to changes in assumptions or changes in the parameters of the model?
4) Have there been other risk assessments conducted and why are they different from the one now in question?

Management
1) Who is responsible for evaluating the data and rendering a decision?
2) What are the legal ramifications of the risk and benefit determinations?
3) Are there factors that constrain the decision such as technical issues, financial limits, time limits, scientific uncertainty, or political ideology?
4) What resources are available for implementing decisions?

Source: National Research Council 1989, pp. 33–37.

adverse conditions for any given activity or technology well in advance. Third, for the sake of argument, let's assume that we can elucidate all adverse conditions and can affix a cost to them. We still need data on their probability of occurrence, data that are sketchy at best. We know, for example, that individuals are generally poor probability estimators (Chapter 2). Given this tendency plus lack of any reliable data, how realistic will these estimates truly be? Fourth, how can we rationally assign a price tag to many of these negative and positive consequences? How much is a life worth? To economists, the value of one life is the value of a person's expected future earnings (Fischhoff, Slovic and Lichtenstein 1979). Obviously, this calculation begs a whole series of ethical questions that are rarely incorporated into the cost-benefit calculations.

Because of the many biases and shortcomings in the cost-benefit technique, it is used less and less frequently in hazards-/risk-management decisions. A newer technique, utilizing many of the basic principles of cost-benefit analysis (without the strong economic bias), has taken over. In risk-benefit analysis risks are compared to anticipated benefits (Crouch and Wilson 1982). Greater risks are tolerated by individuals and society if greater benefits are apparent from that technology or activity. A good example is pesticide use. The risks from ecosystem degradation and human health exposures are generally viewed as acceptable in most countries since the loss of food production to pests poses even greater risks to a hungry world. Unlike cost-benefit analysis, risk-benefit analysis is not always quantitatively driven. While the benefits and risks can be defined numerically, it is often hard to compare data across all risks and benefits. As a result, risk-benefit analysis provides a useful policy tool that weighs many of the risks in relation to the benefits. It creates a balance sheet that often helps to place the project in perspective, rather than producing a defensible number that quantifies the risk to benefit ratio.

Revealed preferences

In reacting to the failings of cost-benefit methods in understanding the social acceptance of technological risks, Starr (1969) proposed his revealed preference method that was briefly described in Chapter 2. He assumed that the historical record of comparing economic risk and benefits would reveal patterns of acceptability across a wide range of technologies. Based on these quantitative comparisons, Starr developed a series of acceptability laws. He concluded, first, that risks are acceptable from a technology if they are proportional to the third power of the benefits; a quantitative determination based on his historical data set. Second, the public is willing to tolerate risks from voluntary activities (on the order of a 1000 times as much) more than from involuntary ones even though both provide the same level of perceived benefit. A good illustration is the public's willing acceptance of smoking (a voluntary activity) and its attendant risks and their unacceptance of red dye, nitrites or other food additives. His third law claims that risk acceptability is inversely related to the number of people at risk; when more people are at risk, the level of acceptance of the technology or activity is much lower. A good example are sugar substitutes such as cyclamates. Heralded as the modern breakthrough in synthetic sweeteners, cyclamates were widely used in the food and beverage industries with strong consumer support. As soon as the risks, in the form of suspected carcinogens, appeared the substance was withdrawn from the market. For Americans, diet products are the rage, so new sweeteners rapidly took the place of the banned cyclamates, including aspartame and saccharin. While there is conflicting evidence in the medical literature, the overall sentiment is that cyclamates pose less health risks than saccharin (Greenberg 1986).

As was the case with cost-benefit analysis, the revealed preference approach also has severe limitations. Two are worth mentioning. First, and foremost, risk is defined as human fatalities. Given that, Starr's method does not include impacts or risks that directly or indirectly affect human health or welfare, environmental degradation, and how these might alter the societal weighing of risks and benefits. He also fails to consider the time dimension. For example, there is often a very long time-lag between technological innovations and the appearance of carcinogens in the environment and in humans, neither of which is reflected in the current market-pricing structures. Second, his method fails to examine distributional issues—who is at risk and who gets the benefits, a very important consideration from our perspective.

Expressed preferences

The expressed preference technique involves an assessment of public views on risks, mostly through direct measurements such as opinion polls. The volatility of public opinion that is influenced by the media coverage of current events contribute to the issue-attention cycle (Downs 1972). Public sentiment is directly tied to specific events creating peaks and valleys in opinion surveys. When the issue is fresh in their minds, such as a chemical accident, public attention is focused and opinion polls reflect high levels of concern. When there is no crisis event, concern levels fall dramatically until the next scare of the week. Because of the changeability of views, expressed preferences have only limited usefulness in public policy decision-making. Needless to say, one of the important findings from this method was not which risks were acceptable but the divergence between expert and non-expert opinions, an issue I discuss later in this chapter.

Risk assessment

Risk assessment uses data to define the health effects of individual or population exposures to hazardous materials and/or situations (National

Research Council 1983). There are four stages in preparing a risk assessment. The first step involves hazard identification or the determination that a particular chemical is linked or not linked to a particular health effect. The second stage examines the relationship between the magnitude of exposure and the probability of occurrence of the adverse health effect identified in step one. This step is often referred to as a dose-response assessment. Next is exposure assessment. In this stage the range and extent of human exposures are calculated either before or after regulatory controls are in place. The last stage is a description of the nature and magnitude of human risk as well as some measure of uncertainty for the estimate. This step is called risk characterization.

As was the case with cost-benefit analysis, there is considerable disagreement over the use of risk assessments in regulatory policy (Clarke 1988a, b; Lave 1989; Daggett, Hazen and Shaw 1989; Stever 1989; Commoner 1989; Finkel 1989). Most of these conflicts centre on scientific issues of measurement, inference and use of quantitative data in public policies. In theory, risk assessments are 'objective' attempts to numerically define the extent of human exposure to chemical, physical or biological agents. Since we know that science is not objective, it is not surprising that scientists often disagree on the interpretations of the quantitative evidence depending on which side they are representing. The question of whether or not the glass is half full or half empty lies at the centre of many debates on risk assessments.

Quantitative risk assessments (QRAs) are frequently used as a tool for policy-making particularly by regulatory agencies such as the US Environmental Protection Agency (Russell and Gruber 1987). The EPA uses risk assessments, for example, in setting priorities within the agency, often in conflict with public perceptions, legislative mandates, and broader social goals of environmental protection. Another application is for rule-making and the design of regulations. Quantitative risk assessments help to select not only individual targets for control, but the appropriate level of control as well. They are also used to adjust national standards to account for variations in the magnitude and distribution of risks. Finally, risk assessments are frequently used by regulatory agencies on a site-specific basis to facilitate management decisions regarding the nature of contamination, sensitivity of the environment, and remediation alternatives (Russell and Gruber 1987). Site-specific risk assessments are most frequently used for hazardous-waste sites to judge their severity and priority placement on either Superfund or state remediation programmes.

Despite these attempts to quantitatively determine the value of a life, or the increased risk of dying from cancer, we still have rather ineffectual tools for making societal risk and hazard decisions. While we can place a numeric value on life, many would argue that public policies made solely on these grounds are extremely short-sighted and reflect anthropocentric world views. Science has not given us these dilemmas and will rarely solve them. The assessment of risk is a social choice and reflective of broader values and ethics such as ecosystem functioning, planet stability, and reductions in materialism. All of these are needed to produce acceptability decisions. Discussions on environmental risks as well as human health risks are necessary to truly reflect the value of a life, be it yours, mine or the great blue heron's in a nearby lake.

Identifying risks and hazards

There are two main avenues for identifying risks and hazards. The first approach centres on the identification of human health risks that are used in the quantitative risk assessments described above. The second approach involves the identification of hazards. For our purposes, this aspect is more important since it forms the basis for geographical research into technological hazards. Unlike quantitative risk assessments, hazard identification is not constrained by a singular focus on human health impacts. It is a more comprehensive view that involves environmental health including degradation arising from technological failures or the use of complex technologies.

Risks to human health

Normally, human health risks are based on involuntary exposures to hazardous substances. The most common methods for assessing these risks are disease clusters, epidemiologic data, animal-bioassay data or experimental studies of carcinogenic or mutagenic properties (National Research Council 1983; Lave 1982; Greenberg 1986). Let's begin with disease clusters since they are the oldest and most widely used method. Case or disease clusters are based on an abnormal pattern of disease that is identified. There are a number of very famous cases (Love Canal and Woburn, Massachusetts, come to mind) involving cancer clusters and their relationship to chemical contamination of the environment. This method, however, rarely provides the conclusive evidence required by regulatory agencies. It is hard to judge whether the cases are merely statistical anomalies since the population at risk is rarely known (Wartenberg and Greenberg 1990). The case cluster method is an extremely useful technique in initially raising concern about environmental contamination and its relation to human health. But, it necessitates a more detailed and thorough epidemiological study to adequately determine cause and effect.

Perhaps the most accepted evidence of human health risks are field studies that show an actual connection between an environmental contaminant and human disease. The evidence is often very hard to get, since there are so many chemicals in the environment, and it is hard to isolate the one that might cause the disease. Also the time-lag between exposure and appearance of the disease ranges anywhere from immediately (acute exposures) to decades (chronic exposures) for some cancers to develop. Because of the costs of large scale epidemiological studies, this method for identifying health risks is not used as often as others.

Bioassay data form the foundation of toxicological research. The operating premise is that results from these animal experiments can be extrapolated to humans. There are, however, a number of questions that lead to disagreements over the results. For example, what species and sex of that species should be used in the assay? What degree of statistical confirmation is necessary for a positive result? How should evidence of different metabolic pathways and rates between animals and humans be included in the assessment? As Greenberg (1986) states, 'Whether the data come from human studies or bioassays, the most controversial phase of the process is extrapolating dose and response relationships to people. This is because almost all of the data about doses and responses are based on responses to high doses. The low dose effects are extrapolated from high dose data' (p. 5).

There is considerable debate among cancer researchers, many who think the risks of synthetic chemicals have been over-estimated because of design flaws in the bioassays used to calculate carcinogenesis. One noted researcher has gone so far as to suggest that chemicals naturally occurring in foods are a far more significant source of carcinogens than synthetic chemicals such as 'Alar' (Marx 1990).

Another approach in identifying health risks are short-term experimental studies that test for carcinogens and mutagens. Because these tests are quick and quite inexpensive they are often used to initially screen chemicals and identify carcinogens. However, these tests are rarely conclusive in and of themselves. Additional bioassay or epidemiological analyses must follow to corroborate the initial suspicions of the carcinogenic or mutagenic properties of the chemicals.

The last health risk assessment method involves comparisons of molecular structure and searching for similarities in chemical structure that might identify carcinogens. Lave (1982) provides the example of coal tars, many of which are human carcinogens. The reasoning goes that if some of them are carcinogenic, then one might logically conclude that most are given similarities in chemical structure. Again, this technique is most appropriate as a screening device to develop a hierarchy or priority setting for further testing using the aforementioned methods.

Environmental risks

Identifying environmental risks is complex because of the dual nature of the task—individual human health risks are important as are ecosystem impacts. Environmental risks arise from a number of sources including technological systems failures, environmental processes, human behaviour or some combination of these. Techniques for measuring these risks and impacts fall into three broad categories: monitoring and surveillance; modelling; and screening and testing for exposure/response (Whyte and Burton 1980). Table 3.2 illustrates the relationship between these sources of environmental risk and selected techniques for assessing them.

Table 3.2 Selected techniques for defining environmental risks.

	Source of risk		
Approach	Technology	Environment	Human behaviour
Monitoring/ surveillance	Pollution emissions	Ambient quality GEMS EOS	Epidemiological surveys
Modelling	Event trees	Physical transport	Migration
	Fault tree analyses	GCMs	
Screening/ testing	Product tests	Persistence	Acute toxicity
		Recovery rates	Lethal dose 50
		Exposure/ response	Lethal concentration 50

Source: After Whyte and Burton 1980.
GEMS = Global Environmental Monitoring System; EOS = Earth Observation System; GCMs = General Circulation Models.

Environmental surveillance

Observations on the spatial and temporal patterns of environmental indicators as well as variations in these patterns are absolutely crucial to our understanding of environmental risks. This is why environmental monitoring programmes are so essential at all geographic scales. The Global Environmental Monitoring System (GEMS), for example, provides comparative data from many world regions and is used to assess the effects of transboundary pollutants such as acid rain, increased concentrations of contaminants in the air such as CFCs, marine oil spills, and water quality. Monitoring programmes like these provide baseline data that can be examined both historically and geographically. This enables a comparative assessment of environmental quality at the global or regional scale. Annual statistical reports such as those published by the United Nations and World Resources Institute (UNEP 1991; World Resources Institute 1992) rely on such environmental monitoring systems. Smaller-scale environmental surveillance or monitoring programmes are available at regional, national or local levels. Many of these are legislatively driven, having been established to monitor the effectiveness of pollution clean-up programmes.

Causality

The second stage in environmental risk assessments is to delineate the chain of events, pathways, and processes that connect the cause or source of the problem to its effect(s). Environmental models are often used to examine these linkages. These models take many forms, ranging from schematic diagrams such as food chains, to purely mathematical representations, to more sophisticated and interactive computer models of environmental processes and responses. At the global level, General Circulation Models (GCMs) are widely used to develop scenario-based estimates of global warming and the locations of anticipated impacts. While far from perfect, the use of models helps to simplify the complexity of many of these problems, so we can gain a better understanding of the processes at work, how they interact, and the likely responses.

Exposure response

A final stage in the environmental risk identification is to determine the relative characteristics of the impact or exposure (dose) to the types of environmental responses (effects). This last link helps to codify the relationships between contaminants and environmental processes and provides opportunities for intervention in the risk system. Dose-effect relationships are illustrated using curves and are most useful in predicting the consequences of either very high or very low exposures; those that are extremely difficult to observe or otherwise measure. Two main types of dose-effect relationships are worth noting (Whyte and Burton 1980). The first is called 'threshold relationships' (Fig. 3.1a) where exposures must meet some threshold value before the effects are noticed. In this schematic, low exposures result in no risk until a certain level of exposure is reached resulting in a gradual increase in effects (curve 1). A variant on this is illustrated in curve 2, where low exposures are tolerable with no effect, but the cumulative nature of the exposures takes off at some point where the effects increase dramatically. An example, is the role of toxins in the environment that can initially be sustained at low levels for a period of time. As the toxicity in the soil, for example, increases, then a threshold is reached where adverse effects are noticed such as declining fertility and loss of crops.

The second dose-effect relationship is called 'zero-threshold' and means that zero risk occurs only at zero exposure (Fig. 3.1b). The classic linear association between dose and effect is illustrated in curve 3. It assumes that there are no safe exposures. Often called the *de minimis* principle, this model of dose-response relationships is widely used by regulatory agencies such as the EPA or Food and Drug Administration, much to the

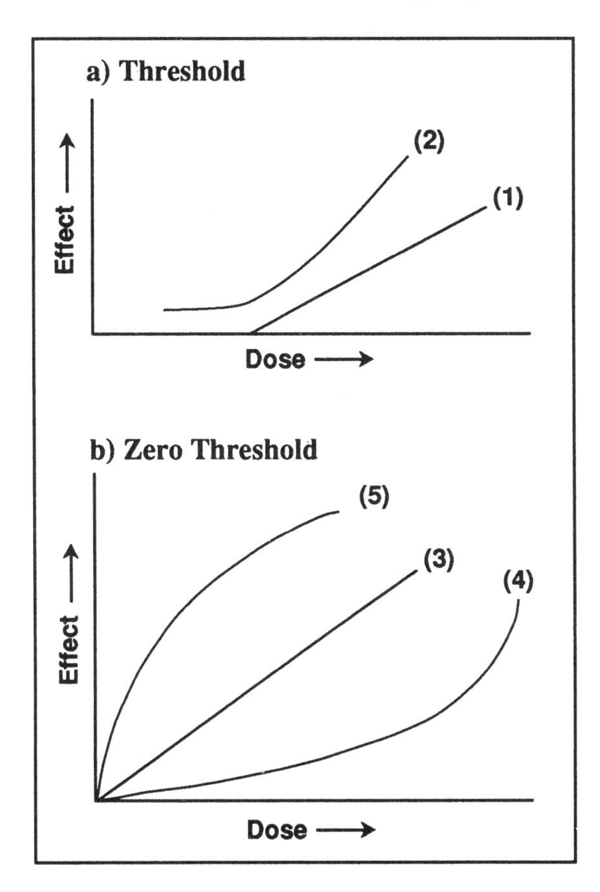

Fig. 3.1 Dose-effect relationships. Threshold relationships (a) include those where low-level exposures are tolerated and accumulate until a threshold is reached before the effects are noticed. Zero-threshold relationships (b) are exposure-dependent; that is zero risk only occurs at zero exposure.

chagrin of industry and some scientists. Curve 4 shows a variant where lower sensitivity of risk is related to lower exposures. Conversely, curve 5 highlights greater environmental sensitivity to lower levels of exposure.

Once we have some basic information on the range, extent, and severity of environmental risks, how is this knowledge incorporated into national and international policies? Policies involving choices among risky alternatives or risk-risk situations (experimental drug use or not) are much easier to manage (Lave 1987). Choices involving how safe situations (nuclear power *versus* fossil fuel use) involve the selection of how much is to be sacrificed for the increased safety—an explicit social choice.

Clearly, the urgency or immediacy of the risk lends itself to differential assessments. If the risk clearly poses an immediate public threat such as a train derailment spilling toxic materials, immediate mitigation as well as regulatory reform is more likely than with those risks whose effects are more in the future, such as global warming. Crisis response or seat-of-the-pants decision-making is the operative word here. One of the problems of course, is the definition of 'imminent'. This is especially true in instances of chemical regulation, where there is a long time-lag between exposure and environmental contamination, or global warming, where activities occurring today will have effects well into the next century.

Another factor that prompts more immediate policy changes are accidents. The increased regulation of safety in design and training and the implementation of stringent emergency-response plans for nuclear power plants was a direct result of the accident at Three Mile Island in 1979. There are many other factors that facilitate prompt public policy actions (Table 3.3). These are instructive and illustrate the salience and usefulness of risk-assessment methodology in the public policy debates. It also highlights the shortcomings of quantitative risk assessments—their inability to handle uncertainty (Roberts 1990), their failure to include the social implications of risk (Freudenburg 1988; Zeckhauser and Viscusi 1990), and divergent views about risks between the public, experts, and policy-makers. Despite the controversial nature of risk assessments, they have improved environmental decision-making by focusing on the critical scientific uncertainties underlying policy decisions (Goldstein and Greenberg 1991). They are useful in assessing hazards and risks from technology already in place—a more reactive or after-the-fact approach.

Table 3.3 Factors leading to prompt public action.

Risks are judged to be serious and action is more likely to be taken when:

- a serious event or accident has just occurred
- there is a high probability of a dangerous occurrence
- danger is imminent
- the danger is new and unfamiliar
- the danger evokes images of fear unrelated to the risk
- the danger is carcinogenic, mutagenic or teratogenic
- the damage is acute and short term
- the damage is irreversible
- many people are potentially affected
- instances of harm are clustered in time or place
- children are affected
- cause—effect relationships are scientifically understood
- there is direct impact on people
- national security is involved
- it is publicly known
- it has mass media attention and/or political attention
- consequences are highly damaging to economic and trade interests
- alternatives are available

Source: Whyte and Burton 1980, p. 69.

They are less useful, however, in deciding on future directions or primary prevention (the risk or consequence of taking or not taking action). Clearly, they have limited value in enhancing societal choices regarding risks and hazards. As Freudenburg (1988) suggests, 'Although often overlooked, human and social factors play vital roles in technological systems; real-world risks, far from being free of such inconvenient "people factors," are indeed often dominated by them' (p. 48).

Who's right: the experts or the public?

It is clear from the preceding discussion that risk assessment is a useful tool in identifying risks and hazards and in quantifying the benefits and costs of reducing them. But their role (or the role of the risk assessors) in making policy choices about which risks society should bear is less clear cut. Here we come into direct conflict between the importance of science and scientific judgements *versus* public judgements. Who is to say whether scientists or the public have better or more correct information? Perhaps they simply have different information. Whose judgements should be used in making societal choices about risks, hazards, and safety?

Expert *versus* public views

As I illustrated with the Alar example in Chapter 2, there are some significant differences between how the public perceives and responds to risk *vis-à-vis* experts. The controversy over expert *versus* public views on risks is a product of the social changes during the last three decades. American society, in particular, has felt these changes more acutely than most. Americans have become more affluent. They have achieved a comfortable level of economic security resulting in more time to devote to political and social issues. A second major shift is the increasing reliance on technology for many aspects of our daily lives ranging from the food we eat (petrochemicals for the pesticides), to the types of work we do (computers, automated manufacturing systems), to how we spend our free time (video arcades, listening to compact discs). In most cases, these technologies are controlled by large, multinational corporations who exercise an enormous amount of economic and political control and clout.

Another social change is the American public's trust of institutions. Increasingly, the public is distrustful of all institutions, be they power companies, the government or even science. This distrust is partly historical resulting from the strife over the social reforms during the 1960s and 1970s as well as a number of key American political events (the Vietnam War, Watergate, Supreme Court confirmation hearings), and environmental catastrophes or near catastrophes (Three Mile Island, Love Canal, and the Exxon Valdez spill in the Gulf of Alaska). All of these events coupled with general American cynicism about government, big business, the courts, and most other institutions, is reflective in the current American distrust of institutions of all types.

New institutions were created during the 1960s and 1970s to protect the environment and the public's health such as the Environmental Protection Agency or the Occupational Safety and Health Administration. Informing patients, workers, and the general public of risks in the workplace, or in the environment became standard operating procedure. Risk information became more widely disseminated and available to anyone who requested it. As the National research Council (1983) stated 'These changes created new public institutions whose purpose was to make technological decisions in the public arena and that resulted in new settings for conflict' (p. 64).

Not surprisingly the public debates over risk became even more politicized, with concomitant changes in many of our political institutions. Public involvement in US decision-making has a long history, particularly in the environmental arena. It has never, however, had the voice that it now does under the whole series of 'right-to-know' laws (see Chapter 4).

Regulatory reform has occurred at all levels as the 'public interest' is redefined and codified in law. Tort law (legal action to determine who is responsible for an injury) was broadened to allow for greater flexibility in who can seek legal redress and over what time period. This effectively allowed people who were exposed to carcinogens 20 or 30 years ago to sue for damages from the manufacturer of a particular chemical compound in question. Regulatory rule-making has also undergone changes; agencies are now required to provide detailed scientific analyses that support regulatory action, rather than just being able to do it by administrative fiat. Of course, the proposed rules and supporting evidence are open for public comment before they go into effect, further encouraging public discourse.

Perhaps the most profound change in the last 20 years that intensified the debate between experts and the public is the institutionalization of scientific conflict. Governmental decision-making on technological issues is driven by science. In fact, scientific experts and science advisors are becoming so entrenched in policy-making, many are beginning to investigate the role of this fifth branch (Jasanoff 1990). Also, divisions in scientific opinion between experts, often gets played out in public view, making the nightly news or front-page of the newspapers. This has led to a new level of scientific advocacy. Each side of the controversy calls their own litany of scientists to refute the findings and conclusions of the other. As we have seen in Chapter 2, science is a particular way of viewing the world, and not the purveyor of objective truth as many are led to believe. This becomes increasingly clear in technological debates and policy choices about risks and hazards that science does not always provide the best answers on risk acceptability.

Risk communication

As a consequence of our understanding about the divergence in perceptions of risk between the public and experts, and the ensuing debates over the acceptability of such risks, a whole new area of study developed called 'risk communication'. Risk communication is a process that develops and delivers a message from the expert or agency to the public. This one-way flow is designed to enable the public to better understand the risk of a particular option. The assumption is simple, if the public understood the technology and the science of the risk calculations, they would be more accepting of the risks or the technology in question.

A considerable amount of money has been spent by regulators and industry in developing risk-communication 'manuals'—guides to inform the public and the press (Covello, Sandman and Slovic 1988; Covello and Allen 1988; Sandman *et al* 1987; Hadden 1989). While laudatory, these assume that the expert or scientific opinion has more validity in the decision-making process than the public's. However, the definition of risk communication is widening to incorporate a two-way dialogue between regulators or managers and the public (Chess and Hance 1989; Krimsky and Plough 1988; Kasperson and Stallen 1991). Thus, risk communication is now viewed as a process of interactive exchange of information; providing the content of risk messages as well as balance in views and accuracy of the message.

What types of problems might be encountered by both the sources and recipients of risk messages? For starters, lack of credibility of either the message or its source alters the communication process. Recipients' views about the accuracy of the risk message are affected by a number of factors including the discrepancies between perceived fact and the position taken by the message, past history of deceit or misrepresentation of facts by the source (in other words, credibility), self-serving or selective use of information in the message, and contradictory messages from other credible and highly reliable sources (National Research Council 1989). Secondly, messages must be understandable to non-technical people using non-technical language. This does not mean that messages must be watered down. Rather, they should contain issues of uncertainty presented in a form that is easily understood rather than in numeric or probability terms such as an 8–10 chance. Conflict often arises because the source and the recipient are not talking the same language, so to speak.

Other problems that arise in the communication process include the limitations inherent in preparing risk messages under emergency conditions, such as during a nuclear power plant failure, or chemical plant explosion. Relevant information may not be known at the time, yet the public is clamouring for knowledge in order to make some decision about their own behaviour in response to the emergency. Finally, there are many things that compete for our attention, so we may not always pay attention to risk messages. Conversely, we may encounter difficulty in gaining access to information from officials who don't listen or fail to provide adequate information to us in the first place.

All this suggests that risk communication is a tricky business and not entirely successful. There are limitations in the process and how it is used to affect public policy. We can, however, view success in a number of different ways. Risk communication is successful, according to the National Research Council (1989), if it raises the level of understanding of the relevant issues for all sides and all parties are satisfied that they were provided adequate information within the limitations of existing knowledge. We should remember, however, that successful risk communication doesn't necessarily lead to better decisions since it is only one part of the risk-management process. Furthermore, it will not result in a consensus over controversial issues which, more often than not, are clashes of values and competing interests. Finally, while the message from the expert is important, so is the public's view. The expert can provide factual information, but the public provides the values to assess those facts and their acceptance or rejection and thus play a pivotal role in societal decision-making processes.

Role of the media

The media play an important role in amplifying and attenuating the public's perception of risks and are a key linkage in the risk-communication process. The media is confronted with as many challenges about how to portray risk information as are regulators. The media, for the most part, depend on facts and opinions from officials and this tends to bias reporting in a conservative way (Clarke 1988a). More often than not, they are not the primary cause of the risk-communication impasse. The problem lies in the inability of scientists and journalists to communicate to one another and understand the pressures and constraints under which both operate (Sandman *et al* 1987). Scientists often dismiss journalists and refuse to talk in language or sound bites that are easily understood, preferring to stay behind the cloak of scientism. On the other hand, journalists only want the juiciest and most spectacular findings (those that sell to their editors and the public) and fail to recognize or appreciate the technical and social dimensions of the issues (Nimmo and Combs 1985; Greenberg *et al* 1989). Journalists often fail to frame the problem correctly and to present the caveats in the findings, further disenfranchising the scientist. Finally, not all media are alike, and scientists need to recognize the differences between print and electronic, local, regional, national, and international news-gathering organizations (National Research Council 1989).

The global-warming issue provides a good example of many of these themes. It incorporates all the elements we've described in making social policies—divergent perceptions, risk communication, the role of the media in heightening or downplaying risks, and public policies.

Global warming: a case study

Public opinion polls from 1982–1991 have included at least one question on the greenhouse effect or global warming (Dunlap 1991a). In 1982, for example, only 12% of those surveyed felt the greenhouse effect posed a very serious problem. This percentage doubled by 1986, and by 1989 41% of those sampled felt that the greenhouse effect posed very serious problems, while another 34% replied it was a somewhat serious problem. There has also been a change in the public's evaluation of the environmental threat posed by the greenhouse effect. In 1987 45% of those polled felt it posed a clear environmental threat, and two years later this jumped to 69% (Dunlap and Scarce 1991). Despite this elevated concern level, only 39% of the public felt urgent governmental action was required in 1989, and this percentage fell to 34% one year later.

Internationally, there is more public concern about global warming especially in Europe and Japan where 43% of respondents were very concerned about possible climate changes brought about by carbon dioxide emissions (OECD 1991a) compared to 30% of US respondents. What accounts for such differences and shifts in opinion over time? Is it a function of increasing media attention, the uncovering of new facts by scientists, successful risk-communication strategies, or geopolitical changes making global warming more of a salient issue? Is public concern warranted? Finally, what is the relationship between this public concern, scientific judgements, and the continuing policy debates on the issue?

A chronology of greenhouse politics provides the backdrop for a brief discussion in disparities in lay (non-expert) *versus* expert opinion and their roles in public policy formation. Clearly, there is a high degree of uncertainty regarding the existence of global warming and the need to prevent it among the scientific community. A June 1979 report to the US President's Council on Environmental Quality (CEQ) noted the impacts of human activities on the atmosphere that if unabated, would lead to significant warming of the world (Lyman 1990). Other critical junctures in the greenhouse politics are listed in Table 3.4

The media have been instrumental in heightening public awareness of the greenhouse effect, specifically, and its role in global warming more generally. The 1988 drought in North America coupled with intense heat for most of the summer was portrayed in the press as a consequence of global warming. *Time* magazine's recognition of the Planet of the Year in 1988 further intensified the image of planetary catastrophe as a result of human actions. Both *Time* and *Newsweek* featured cover stories in

Table 3.4 A chronology of greenhouse politics.

1979
June: Report to CEQ on potential for significant warming due to human activities.

October: NAS study on greenhouse effect that concludes a doubling of carbon dioxide would increase global temperatures by 2.7–8.1°F.

1980
January: CEQ report urging inclusion of carbon dioxide problem in global and national energy policies.

January: Goddard Institute of Space Studies report on importance of other greenhouse gases (methane, ozone, nitrous oxide, CFCs).

1983
EPA report, *Can We Delay a Greenhouse Warming?* concludes current CO_2 emissions will raise global temperatures by 3.6°F.

NAS report, *Changing Climate*, finds insufficient evidence to change current energy policy.

1985
October: Villach, Austria, scientific conference on global warming sponsored by UNEP and WHO, 29 nations send delegations.

December: International Year of the Greenhouse called for by Sen. Albert Gore (D-TN).

1986
January: WHO and NASA reports on atmospheric ozone.

June: Senate hearings on global warming and ozone hole, request to EPA and OTA to develop policy options for stabilizing greenhouse gases.

June: Congress earmarks extra money to EPA for climate change research.

November: US proposes international phase-out of CFCs.

1987
September: Montreal Protocol signed by 24 nations.

December: Reagan and Gorbachev agree to continue joint studies on global climate change.

1988
January: Reagan signs Global Climate Protection Act, thus requiring him to propose policy responses to Congress.

April: UNEP/WHO report concludes that climate change will outpace the natural systems ability to respond and adapt to changes.

April: US ratifies Montreal Protocol.

June: Gro Harlem Brundtland calls for global convention on greenhouse effect at international conference on 'The Changing Atmosphere: Implications for Global Security'. Conference also calls for 20% reduction in CO_2 emissions by 2005.

November: Intergovernmental Panel on Climate Change (IPCC) established.

1989
January: British scientists concur with warming predictions.

March: Margaret Thatcher calls for strengthening of Montreal Protocol.

May: George Bush endorses support of a climate convention and invites IPCC delegates to Washington.

May: Helsinki Declaration at UNEP ozone-layer meeting, delegates agree on Montreal Protocol revisions and total phase-out of CFCs.

May: UNEP meeting in Nairobi lays groundwork for climate-change convention.

July: 15-nation economic summit declares urgency of limiting greenhouse gases.

September: Japanese government holds world conference focusing on climate change and deforestation.

October: IPCC delegates discuss convention framework.

December: President Bush proposes US to host first negotiating session on Climate Convention.

1990
Autumn: UN General Assembly establishes intergovernmental negotiating committee for climate-change convention.

Autumn: IPCC report, *Climate Change: The IPCC Scientific Assessment*, published.

1991
February: First negotiations on Climate Convention in Chantilly, Virginia.

June–September: Preparatory international meetings on the contents of the Climate Convention.

Autumn: US opposes emissions reduction targets and financial assistance to developing countries provisions in Climate Convention.

1992
January–June: Preparatory meetings continue negotiating climate-change treaty.

June: Climate Change Treaty signed by 150 nations during UNCED meetings in Rio de Janeiro. Global warming convention asceded to US demands and set recommended levels for curbing emissions of greenhouse gases, but no specific targets.

Sources: Lyman 1990; Schneider 1989b; Stevens 1992.

1988 on global warming from the greenhouse effect and ozone depletion from industrial chemical use. In 1992 *Time* again had cover stories on ozone depletion and the Rio Earth Summit, as did many other news weeklies.

While the greenhouse effect was well known and strongly debated in scientific circles, it wasn't until the end of the 1980s, the warmest decade ever, that media, public, and governmental concern coalesced and greenhouse politics born. All of a sudden, the scientific debates with their attendant uncertainty on the global warming signal (Schneider 1989a) became front-page news, no longer relegated to academic journals; the scientific debates were played out in public. The consequences of global warming and what to do about it were contentious (USEPA 1989b). Everyone joined the global-warming tug-of-war. Those arguing for more research to assess the risks before any action was taken were on one side. Others argued that it was too late for studies, that action was needed now. Even editorial cartoonists got caught up in the controversy lampooning the supposed effects of the greenhouse effect, their location and timing, and the level of scientific uncertainty about the likelihood of impending change (Fig. 3.2).

One of the biggest obstacles to global-warming policies at the moment is the divergence between public and scientific perspectives and policy-makers on the issue. In a recent study using ethnographic techniques, Kempton (1991) examined perceptions of the causes and consequences of the greenhouse effect. He found considerable divergence between public views on the causes of the greenhouse effect and those of scientists. He also found differences in concern levels. For example, according to the public the causes of the greenhouse effect are aerosol spray cans, the ozone hole, cutting trees, and air pollution. Note that none of these represents the chief cause according to scientists: the release of sequestered carbon through fossil fuel burning or deforestation. This cause, could be inferred, however, from the cutting trees and air-pollution responses of the public, but it is unclear whether this link in comprehension, in

Fig. 3.2 Impacts of global warming. Uncertainty about the timing and nature of the impacts of global warming lead to public scepticism and accusations that the scientific community is crying wolf once too often. Source: USEPA 1989b:p. 30.

fact, has been made. The most important issues of lay concern are the lasting legacy to our children who will bear the consequences, sea-level rise, warmer weather in summer, depletion of atmospheric oxygen, and breathing of greenhouse gases. Scientists are more concerned about the profound nature and unpredictability of effects, the rapidity of occurrence, biological disruptions, agricultural shifts, and possible sea-level rise. The only commonality is the concern over sea-level rise.

If we compare expert and public views on the scientific evidence for global warming and the effectiveness of public policies to minimize the impacts, more differences emerge. As can be seen in Table 3.5, the public and the experts disagree on the basic evidence supporting global warming. Not surprisingly, there are also conflicting views on the effectiveness of various policy alternatives. The public thinks banning aerosol spray cans is an extremely important policy option, despite the fact that sprays containing CFCs have been banned in the US for two decades. Energy efficiency, deemed most important by the experts, is viewed as irrelevant by the public. While Kempton's study is far from definitive and should not be viewed as representative of a national sample, it nevertheless provides a much needed perspective. Its contribution is on how individuals perceive and define the risks and evaluate policies to reduce the risks of global climate change. Kempton's findings help to explain the public acceptance of US policy options that place tree-planting above energy efficiency and carbon taxes, an unfathomable and tragic choice as far as the scientific community is concerned.

Table 3.5 Comparisons of scientific and lay views on the greenhouse effect.

	Scientist perspective	Lay perspective
Scientific evidence		
CO_2 increase	Clear evidence	Not known
CFC increase	Yes	Yes
More extreme weather	No	Already apparent
Temperature increases	Maybe	Warmer winters already here
Policy effectiveness		
Energy efficiency	Yes, important	No, irrelevant
Stop using aerosol cans	Irrelevant in US	Yes, important
Halt deforestation	Helpful	Extremely important
Reforestation	Limited potential	Yes, Important
Stricter pollution controls	Irrelevant	Yes
Carbon tax	Yes	Ineffective and unfair
Adaptation without prevention	May be cost-effective	No, avoid decisions by politicians

Source: Kempton 1991, p. 206.

Not only do the public and scientists disagree, but there are disagreements within the scientific community as well. A quick glimpse of global-change research funding illustrates the schism between scientific perspectives. In FY 1990, for example, US $653 million was appropriated specifically for global-change research by the US government. Of that total only US $4.8 million was earmarked for the study of human interactions—the relationship between biologic, atmospheric, hydrospheric, and terrestrial changes and human activities that either stimulate or mediate these changes (Table 3.6). While funding for the human dimensions increased in FY 1993 to US $25.9 million, this still represents a minuscule fraction (1.9%) of total global-change research dollars. This is quite a paltry sum considering the centrality of people in creating the risk and those most likely to be effected by it. So you see that even in the research community, people are still arguing on the existence of risk, probability of occurrence, and its potential impacts.

Table 3.6 US global environmental change research budget (US $ million).

Programme	1990	1991	1992	1993*
Climate and hydrologic systems	291.7	450.7	505.5	629.4
Biogeochemical dynamics	198.7	249.1	288.8	333.2
Ecological systems and dynamics	90.2	140.0	152.8	240.4
Earth system history	7.7	18.2	19.3	23.4
Solid earth processes	57.4	53.6	108.6	105.2
Solar influences	8.8	13.8	18.8	15.1
Human interactions	4.8	28.3	16.8	25.9
Total	659.3	953.7	1109.8	1372.4

* Proposed FY 1993 budget.
Source: Committee on Earth Science 1990, 1991, 1992.

A US National Academy of Sciences panel (1991) concluded that greenhouse warming poses enough of a potential threat to warrant action immediately. They proposed a series of short-term mitigation measures to reduce US greenhouse gases, and initiated a number of adaptation strategies in preparation for a warmer world (Table 3.7). In case of this long-term risk, it may be prudent to believe in the mere possibility and err on the side of caution (IPCC 1990, 1991). The policy-makers at least at this point, are more sanguine, calling for more research, and adopting a very conservative wait-and-see attitude, despite overall public and scientific concern about the issue. Perhaps this is a harbinger of the future where the public and scientists are more consistent in risk and hazard identification and the policy-makers lag behind. Despite the overwhelming public support and scientific recommendations, the US did not support a strong climate-change treaty. At the US's insistence, no specific timetables for carbon dioxide emissions were included other than very general statements about reaching a goal of reductions to 1990 emissions level. The US did sign the weakened climate-change treaty during the Earth Summit in Rio de Janeiro in June 1992, although it is unclear at this time how this will affect US policy.

Table 3.7 Policy recommendations for greenhouse warming.
Reduce or offset greenhouse gas emissions
• phase-out of CFCs
• study social cost-pricing of energy
• enhanced conservation and energy efficiency
• greenhouse warming as a factor in energy-source selection
• reduce global deforestation
• domestic reforestation programme
Enhance adaptation to greenhouse warming
• continue agricultural research for adaptation strategies
• improve water availability through efficiency and better management of existing supplies
• plan for infrastructure to withstand climate changes
• slow losses of biodiversity
Knowledge improvements
• continue data collection and dissemination
• improve weather forecasts
• continue identification of climate-change mechanisms and climate-change models
• research on CO_2 enrichment and how this affects biodiversity
• strengthen research on social and economic aspects of greenhouse warming
Evaluate geoengineering options
Exercise international leadership
• assume full participation in international population-growth programmes including financial assistance
• participate in all international agreements and scientific exchanges on greenhouse warming

Source: National Academy of Engineering 1991.

The regulator's dilemma

When faced with scientific uncertainty and the view that the public is ignorant about risks, how should regulatory agencies respond, particularly when their missions are to protect public health and welfare? How should these agencies proceed to insure our health and welfare? Historically, environmental risks were managed on a pollutant by pollutant basis or by medium (e.g. air pollution, water pollution), or some combination (air toxics, toxic chemicals in water). Rarely were there efforts to determine broader environmental-quality goals or to determine the relative importance of individual pollutants or media to broader environmental-quality goals, although this is slowly changing. Furthermore, little effort was placed on determining the cumulative or synergistic effects of individual contaminants on the entire ecosystem, or on overall human health and well-being. Fragmentation of policies and crisis management governed regulatory responses to environmental risk throughout the 1980s and early 1990s.

Yardsticks and comparable risk

Risks need to be measured against something or compared to something in order to be meaningful. Depending on the analytical technique used, risk comparisons can produce very different conclusions on the relative magnitude of the risks under scrutiny. There are a number of methods used to compare risks.

One of the most common techniques compares a particular risk to natural background levels in order to find the elevated risk level caused by the technology or activity in question. Natural background levels are used extensively to measure the additional risk of radiation from nuclear power plant accidents. This has led to such statements as 'your radiation exposure from a nuclear power plant is less than what you get near your home'. Natural background levels are also used in many pollution standards, providing the baseline for ambient quality measurements.

Another method for comparing risks is to balance one particular risk against the risks of alternatives such as other products, processes, or courses of action. This balanced approach works well when there are, in fact, alternatives such as the substitution of one pesticide for another. It works less well when alternatives are not that apparent.

A third approach is the risk-benefit analysis that we have already described in this chapter. Finally, the last method, comparative risks, simply compares the particular risk in question to other risks to measure its relative importance or relative harm. In this method, the benefits are generally ignored, and the risks reduced to a common metric, usually the number of deaths attributed to the technology or activity. Statements such as 'your risk from dying from cigarette smoking is more than your risk of dying from a train or bus accident' are products of this method.

Within regulatory frameworks, natural background levels are normally used for standard setting. Balanced risk, and risk-benefit analysis are largely policy tools. Comparative risk is now being used to determine priorities for action within the regulatory arena: a clear recognition that some risks pose greater danger than others and that limited resources must be used more efficiently.

EPA's relative risk strategy

In 1978 a Science Advisory Board (SAB) was appointed to provide the US EPA with advice on a wide range of scientific and policy issues. Environmental policy was extremely fragmented relying on laws and programmes that were pollutant-specific using tools that emphasized the end of the pipe controls or remediation. The EPA simply enforced the litany of pollution laws passed by Congress and had very little flexibility in reallocating resources between programmes. This partially restricted their ability to set long-range environmental protection goals. It also hampered their ability to respond to the crisis at the moment, or scare of the week. Reducing the risks or minimizing wastes was not part of the policy strategies.

Starting in 1986, the EPA on the advice of its Science Advisory Board began a self-study on relative risk, trying to assess and compare the risks associated with a wide range of environmental problems. Thirty-one problem areas were identified and compared along four dimensions: human cancer risk, non-cancer human health risk, ecological risk, and welfare risk. This study, titled *Unfinished Business* (USEPA 1987) set the stage for a realignment of agency priorities (Morgenstern and Sessions 1988). One of the most important findings of this internal report was that EPA's priorities were more closely aligned with the public's perception of risks than scientific assessments of it. To enhance the credibility of this report and garner Congressional support, the new EPA administrator, William Reilly, asked EPA's Science Advisory Board to review the internal report on relative risk, evaluate its findings, and develop recommendations for reducing risks. The SAB was also asked to review policy options that linked risk reduction to social and individual choice mechanisms including such measures as economic incentives, laws, or regulations that would shape individual and social choices. In other words, there was a change of emphasis at the agency; a movement away from pollution control to pollution reduction, from risk control to risk reduction.

The SAB report included comparative risk analyses that were developed to facilitate a discussion of disparate environmental problems using a common language (USEPA 1990b). The report concluded that habitat alteration and destruction, species extinction and loss of biological, diversity, stratospheric ozone depletion, and global climate change had the highest relative risk and greatest potential for adverse impacts on the environment and human welfare. It is interesting to note, that only ozone depletion and global warming appeared on the original list of 31 problems initially identified by EPA. Based on human health criteria, ambient air pollution, worker exposure to industrial and agricultural chemicals, indoor pollutants, and pollutants in drinking water were evaluated as posing the highest relative risk. Relative low-risk problems included oil spills, groundwater pollution, radionuclides, acid runoff to surface waters, and thermal pollution.

There are two critical issues implied by these relative risk rankings. First, the natural environment and its degradation has just as much primacy in risk determinations as does the impairment of human health. Second, this exercise illustrated just how inconsistent public rankings of environmental problems and scientific judgements are. For example, in the public's ranking of the top four most serious environmental problems (actively used hazardous-waste sites, abandoned hazardous-waste sites, water pollution from industrial waste, and worker exposure to toxic chemicals) only the last problem poses a high risk according to the EPA report. Given this, how should the regulator respond? Should they go with the scientists or determine their priorities based on public concern knowing full well the public opinion is likely to change based on media reports of the scare of the week? Is comparative risk analysis the most appropriate mechanism for determining environmental priorities?

Recently, the use of comparative risk analysis in setting priorities for the agency was debated among scientists and policy-makers (USEPA 1991a). A sampling of views is found in Table 3.8.

Table 3.8 Contrasting views of EPA's relative risk strategy.

'. . . if we rely solely on scientific assessment of relative risk to set environmental priorities, the public is left out of the equation' (Reding 1991, p. 27).
'. . . a budget process that focuses on those problems that already exist because their risk can be quantified may well do so at the expense of support of primary prevention approaches' (Goldstein 1991, p. 23).
'. . . estimating risk is a process for summarizing science to support decision-making' (North 1991, p. 31).
'However, even a flawless risk measurement cannot define "acceptable" risk' (Burke 1991, p. 39).
'. . . basing environmental priorities on relative risk will quickly lapse into environmental triage' (Mott 1991, p. 21).

Source: USEPA 1991a, pp. 17–39.

How safe is safe enough?

While providing the focal point for our discussion, this question, how safe is safe enough? should not dominate debates about societal choices about risks. Assessing the probability and magnitude of risks still ignores the social acceptance of them. Risks that have rather low probabilities of occurrence, often garner little public acceptance (radioactive-waste depositories). Rayner and Cantor (1987) suggest a reframing of the question to how fair is safe enough? By doing so, the debate over technological options is defined in social terms—social conflicts over need, trust, and equity—not simply the probability and magnitude of likely events. As you read in Chapter 2, environmental equity and justice is becoming more dominant in risk and policy decisions (USEPA 1992).

Clearly, the debate is far from over and highlights more than ever just how difficult it is to determine how safe (or fair) is safe enough, placing the regulator and society in very difficult and tenuous positions. On the one hand they must balance scientific uncertainty and quantitative assessments about risk, while on the other cope with the fear, anxiety, seeming lack of knowledge, and different value judgements of the public. The results are imperfect and more often than not result in tragic choices.

4

Managing technological hazards

'In contrast to the virtues of the *de minimis* approach that proposed to ignore very low levels of hazard, our society seems to have adopted a *de ignoramus* approach that avoids knowing about many hazards' (Kates, 1986, p. 214).

Determining the sheer presence of risks and hazards and their societal acceptance is extremely contentious, often pitting one group against another (regulators *versus* scientists, regulators *versus* public, media *versus* scientists, public *versus* government). Nowhere is the contested nature of risk more visible than in the management of risks and hazards. Scientific uncertainty, regulation, responsibility, prevention, and equity are all issues that bias and otherwise constrain hazards-management systems. But what is a hazards-management system?

Management systems

Hazards management implies a certain degree of controllability of the outcomes—the ability to differentiate between accident and incident, or the acts of God, acts of nature, or acts of humans. According to Kasperson, Kates and Hohenemser (1985), as an activity, hazards management informs us about hazards, helps us to decide on a course of action, and then follows through by implementing the appropriate control or mitigation strategies.

Components

Most hazards-management systems have the following elements: monitoring, research (risk estimation), legislation, regulation, inspection, enforcement, emergency response, and a continual re-evaluation of the system itself. We have already described some aspects of this system, e.g. monitoring and research (or how risks and hazards are identified), in Chapter 2.

Standard setting, another element in the management system, is a legislative or regulatory-driven mechanism that codifies what is socially acceptable. It is a quantity of exposure that is deemed 'safe' or tolerable and is measured using a single value or a range of values with specified high and low end-points. We have standards for a wide range of technologies and products (Table 4.1). Yet, conflicts continually arise over the value that is selected creating adversarial positions: industry prefers ranges or grey values, while health, labour, regulatory, and the public prefer a single numeric or a clear accept/reject value (Whyte and Burton 1980). What is unique and important about standards is that they are legally enforced. Standards violations result in punitive fines, civil and criminal action, or product recall.

Table 4.1 Common standards.

Type	Example
Human exposure to risk	Radiation exposure
Effluent	Stack emissions
Ambient environmental quality	Tropospheric ozone level, drinking water standards
Occupational conditions	Safety goggles and helmets
Product, technology, or process design	Industrial machines
Product composition	Food additives
Product or technological performance	Building codes
Product label/advertising	Fireproof or fire retardant
Product packaging	Child-proof drugs

Sources: Whyte and Burton 1986, p. 116; Lowrance 1976.

A less formal approach are guidelines, which are recommended levels but are not legally enforceable. Often, regulators prefer guidelines to standards because of high levels of scientific uncertainty, disputes over the scientific evidence, or governmental unwillingness to enforce tougher standards. Criteria are a more watered-down version of guidelines that reflect the current status of knowledge about environmental conditions and risks and their effects on humans and the biosphere. Criteria can be upgraded to guidelines; guidelines upgraded into standards as the scientific evidence becomes less contentious. Similarly, guidelines can become standards when the political framework for regulatory control changes especially when there is an ideological shift resulting from a new government created by Presidential or Parliamentary elections.

Once regulations and standards are in place they must be enforced. Regular reporting and inspection programmes are designed to do just that. Health and safety inspections normally occur on a regular basis and are designed to monitor compliance with a whole range of occupational health, public health, and safety regulations. Examples are the cleanliness in restaurant kitchens, peeling paint or insulation in day-care centres, both of which pose a potential threat to patrons in the restaurant or children in the day-care centre. Another example of enforcement is the permitting process. The US National Pollutant Discharge Elimination System (NPDES) issues permits to industries and municipalities under the Clean Water Act that allow them to discharge regulated quantities of pollutants into local water bodies. The effluent is periodically checked both for its composition as well as the volume of material discharged. If an industry or municipality violates its NPDES permit which specifies the type and quantity of material that can be discharged, it is then held in violation of the Clean Water Act. Normally the offending party argues to pay a fine and to comply with permit requirements thereafter.

Emergency response is another element in the hazard-management system and only comes into play during accidents or incidents. According to Kates (1978) accidents are distinguished from incidents, the former involving an unintended direct failure of a device, technology, or a technique and its practice; the latter a threat posed by some activity unrelated to the threat itself. A good example of the distinction involves the chemical industry. A failure in a valve at a manufacturing facility resulting in an off-site release of chlorine would be termed an accident. If, on the other hand, a train derailed, puncturing a tank car carrying chlorine allowing it to vaporize, then we would call this an incident. While the distinctions are subtle, they are extremely important in determining responsibility and ultimately liability for damages.

Risk managers

Throughout the risk-management system there are individuals and agencies that coordinate, assess, and regulate risks and hazards. I shall call these the risk managers and they involve many types of people: administrators, interest groups, decision-makers (a very generic term), planners, social agents and even scientists. The explanatory scheme by Kates (1978) is useful since he distinguishes hazard managers by the differing roles they play in the management process: hazard-makers, risk-takers, guardians, and assessors. The roles periodically change depending on the issue and thus public agencies may find themselves as assessors one day, and risk-takers the next.

Hazard-makers intentionally create hazards. The military and hostile governments may intentionally create environmental hazards during the course of armed conflict such as the Persian Gulf War, or under the guise of protecting national security and national interests. Manufacturers can occasionally become hazard-makers when they willingly accept quality control assurances or safety designs that work 99% of the time. They become a hazard-maker when their product fails that one time out of a hundred. Perhaps more to the point, we are all hazard-makers in our own daily lives. Every time we get into our cars for example, we are creating a potential hazard not only to those travelling with us, but for some unsuspecting soul who happens to be in the wrong place at the wrong time.

Risk-takers are another broad category of roles in the hazard-management system. Whether at the individual, group, or societal level, we all engage in some type of risky endeavour; we take calculated risks. The distinction between voluntary and involuntary exposures (see Chapter 2) help to explain this position. We are more likely to be risk-takers when we engage in voluntary activities such as smoking or automobile use. As individuals we are more likely to be risk-averse when we are involuntarily exposed to risks and hazards such as nuclear power, pesticide residues on food, or biotechnology. Despite our knowledge about risks and hazards, it is still extremely difficult to distinguish and justify risk-taking or risk-averse actions on the part of individuals, agencies, or society as a whole.

Guardians protect against risk. At the individual level, guardians are parents who protect children, or doctors who protect patients. At the community level, we have guardians who protect our safety (police), respond to emergencies (fire and medical personnel), and insure healthy drinking water or garbage pickup (local government). In the risk-management system, guardians are the regulators and enforcers who partially determine acceptable levels of risk, but whose primary role is to ensure safety.

Finally, we have the risk assessors: the professional élite that have emerged to provide the 'science' to the study of risk. These risk professionals (Dietz and Rycroft 1987) are involved at almost every level in the risk-management system—research, consulting, regulation, emergency response, policy evaluation. The strength and pervasiveness of this new élite and their non-regulated role in risk management has led many to question the role of science advisors (including the risk professionals) in public policy-making (Jasanoff 1990).

Constraints on management systems

There are many issues that constrain hazards management and pose significant challenges to the hazards manager. Kates (1986) suggests a number of possible limitations on any hazards-management system. One is our inability to anticipate and prevent a catastrophic accident (the surprise factor). Similarly we can't prevent all hazards nor can we reduce risks to zero. There are also institutional limits (social and political contexts). Finally, there are moral choices (what I called tragic choices in Chapter 3) with inherent conflicts between efficiency and equity. I will describe a few additional constraints that are noteworthy as well.

Scientific uncertainty

While the level of scientific uncertainty in estimating risks and hazards posed problems in making acceptable risk decisions, it creates even greater dilemmas for the hazards manager. The tragic choices found in risk-acceptability judgements are passed onto the management system. As we shall see, management options sometimes have their own tragic choices. Often regulations are promulgated based on incomplete or contradictory evidence, largely encouraged by public fears or the surprise factor (Kates 1986). The case of oncogenic (cancer) risk from pesticide residues in food is a good example of contradictory evidence.

Cancer, the de minimis *principle, and the Delaney paradox*

Cancer and cancer-causing substances generate a high degree of fear and public concern in many parts of the world. This is most evident in the United States with its concern over pesticide use on agricultural crops. Pesticide residues in food have a long-standing tradition of public concern dating back to the publication of Rachel Carson's influential book, *Silent Spring* in 1962.

The conflicted nature of pesticide control is evident in the differing legislative and regulatory approaches. On the one hand, pesticides are useful so long as they have no side effects that harm humans or the environment. On the other hand, trade-offs must be made to balance the use of pesticides and their risks *versus* the benefits of a stable food supply (Fig. 4.1). The Environmental Protection Agency is the guardian of pesticide safety in the US. On occasion, it also functions as the risk assessor in the management system. The system was established by legislative mandate, first by the Federal Insecticide, Fungicide, Rodenticide Act (FIFRA) of 1947 and subsequent amendments, and then by provisions in the 1954 Food, Drug

Fig. 4.1 Balancing risks and benefits. Balancing risk and benefits confounds the management process since not everyone is in agreement on the trade-offs. Used with special permission of North America Syndicate, Inc.

and Cosmetic (FDC) Act (National Research Council 1987).

Under FIFRA, the USEPA is responsible for regulating and approving pesticide use according to cost-benefit calculations. EPA's role under the FDC Act is a little different. Here it must set the legal allowable limits for pesticide residues in foods. Under Section 408 tolerances are established for pesticide residues on raw commodities such as fresh fruit and vegetables. The tolerances are to protect human health, but the EPA must also consider the need for an adequate, economical, wholesome food supply in establishing the tolerance levels. Implicit in the setting of tolerances or standards is a risk/benefit calculation.

Section 409 of the Food, Drug, and Cosmetics Act complicates EPA's role in standard setting. This section governs pesticide residues in processed food such as tomato sauce, or canned fruit cocktail. Additives and pesticide residues must not harm individuals. Rather than using the risk/benefit procedure, a zero-based risk level is required. That is, acceptable risk is defined as zero risk. More than 7000 pesticide tolerances have been established to date, the vast majority on raw commodities (Section 408) (Table 4.2).

Table 4.2 Number of Section 408 and 409 food tolerances.

Pesticide type	Section 408 (raw commodities)	Section 409 (processed foods)
Insecticides	3654	63
Herbicides	2462	39
Fungicides	1256	20
Total	7372	122

Source: National Research Council 1987, p. 19.

One of the most controversial aspects of the FDC Act is the infamous Delaney Clause which governs food additives and pesticide residues in processed food. The Delaney Clause, a zero-risk standard, prohibits any food additive or pesticide residue in processed food that induces cancer (either benign or malignant tumours) in either humans or animals. Furthermore, the Delaney Clause expressly prohibits the EPA from establishing tolerances for any processed foods for any oncogenic pesticide even if the benefits clearly outweigh the negligible risk. If some scientific study found the pesticide or food additive caused non-cancerous tumours in animals, this alone precludes its use in processed food. Since pesticide residues become highly concentrated during processing, this is a very cautious approach. On the other hand, where the risks are truly negligible some consideration of the benefits of use might overshadow the decision.

The use of the fungicide benomyl on tomatoes is a good example—it is allowed on fresh tomatoes, but not in processed ones. Benomyl is used extensively in Florida to control fungal diseases on tomatoes. Since 98% of Florida's tomato crop is sold as fresh produce, it is the fungicide of choice. In California, however, where most of the tomato crop is processed into sauce or ketchup, benomyl cannot be used, resulting in crop losses due to the fungal diseases.

In having two masters (FIFRA and FDC Act), so to speak, EPA has a very complicated management system for pesticide residues: one set of rules apply to raw commodities, another to processed food even though the crop is the same (e.g. tomatoes, peanuts). Recently, they have tried to develop a *de minimis* approach to pesticide regulations (Whipple 1986). Using the negligible-risk standard (*de minimis*) rather than the zero-based standard provides management flexibility within the agency. The application of the *de minimis* principle to all additives and pesticide residues on both raw and processed foods could reduce oncogenic risks over time. Since the current policies only consider one pesticide or chemical at a time and are based on the zero-risk standard, it is no wonder that the risk-management process for pesticide residues in food is highly variable and subject to politics and criticism from industry, the public, and the regulators themselves (Asch 1990).

Comparative national risk profiles

Scientific uncertainty also stymies comparative assessments of environmental risks at the national and international levels. One obvious mechanism to improve our understanding of environmental risk and hazards is the development of national risk profiles. First proposed by Whyte and Burton (1980), risk profiles provide a mechanism for identifying and categorizing the panoply of risks that face a nation. A national risk profile, for example, would illustrate the range of risks within the particular society, identify data needs, assess priorities, and evaluate current management systems. The objective is to develop a coherent national policy for risk and hazards management. These national risk profiles would not only examine risks within the nation, disaggregated to local spatial units, but also entail an assessment of that nation's contributions to regional and global risks; how internal activities contribute to regional and global environmental hazards. While a daunting task, these national risk profiles could provide the necessary documentation for the development of cogent response and integrated hazards-management policies at a variety of geographic scales.

Obviously, the content of risk profiles would vary from country to country. However, for comparative purposes, standardized information is necessary. The following is a sample of data suggested for each national profile (Kasperson and Kasperson 1987, p. 115).

1) Risk identification and estimation: risk sources by broad generic categories: natural, environmental degradation, infectious diseases, technology, industrialization, urbanization.
2) Risk events: actuarial evidence on number and frequency of risk events, type and magnitude of consequences.
3) Historical context: historical evidence on trends in risk, its sources, risk events, and risk consequences.
4) Distribution of risks: risks by geographic region, economic sector, social groups, age groups.
5) Population affected: identification and estimates of the population at risk, vulnerable sub-populations such as children, elderly, women, handicapped.
6) Contemporary context: sociopolitical contexts and constraints that amplify or attenuate risks and hazards such as media, democratic, institutions, regulatory control, popular dissent.
7) Data: data quality and comprehensiveness on major risk areas (spatial and temporal coverage), new data requirements.
8) Transboundary/global risks: host-country contributions to transboundary or global risks by source and quantity.

While many assessments of the world's resource base are readily available (World Resources Institute 1990, 1992; UNEP 1991; OECD 1991a, 1991b), national risk profiles do not currently exist in any systematic fashion, despite repeated calls for their creation. Clearly the need to examine the range of risks that threaten human and global existence is important. The risk mosaic of one country helps provide data not only for that nation but enables a more complete and standardized comparable set of information across countries, thereby reducing scientific uncertainty.

The commons and transboundary issues

The issue of responsibility underlies many of the contemporary approaches to hazards management. If the risk occurs in my jurisdiction, I will exercise control over it. If a chemical facility is polluting an adjacent stream, the local or state Department of Environmental Protection intervenes to control the pollution at the source to minimize impacts to local residents and the local environment. But what happens when the risks are generated in one place, but their effects are felt elsewhere? Who is responsible? Furthermore, who is responsible for hazards that affect resources such as the air or the oceans that we all use?

In his classic article, 'The tragedy of the commons', Garrett Hardin (1968) argued that resources for which there was no formal ownership, that is they were owned by everyone (common property resources), lend themselves to rapid exploitation. The discrepancy between ownership and management responsibility enables exploitation since the costs are borne by all the owners of the resource, yet the benefits accrue to the individual. Unless some institutional arrangement is made that prohibits overuse and encourages conservation, the exploitation continues unabated.

Consider for example when a community generates sewage and the sewage is dumped directly into coastal waters (hazard). Those same waters are also used for shellfishing, the primary economic activity for the town. It would be foolhardy for one individual not to dump their sewage into the coastal waters, since they would have to find an alternative method of sewage disposal while still suffering from the pollution of everyone else. In other words, they would not reap any economic benefit from their own pollution reduction; they would still be exposed to the pollution and eating contaminated shellfish.

Common property does not imply open access, since social institutions often impose restrictions on the use of the resources. Common property dilemmas are usually solved by governmental intervention or privatization, both of which have been known to exacerbate the 'tragedy' (McCay and Acheson 1987).

Transboundary hazards provide similar management problems and constraints. National boundaries are permeable to most environmental risks; the transport of pollutants by air or water knows no nationality nor do pollutants adhere to political delineations of space. For example, while acknowledging that the US contributes to Canada's acid-rain problem, the exact nature of the contribution in terms of a specific cause (e.g., Industry A's smokestack emissions and acidification of Lake Wobegon) is difficult if not impossible to prove. So who is responsible and who should take the initiative to manage or reduce these risks—the US, Canada who also contributes, or both? The transboundary issue becomes even more problematic with the failures of large and complex technologies such as nuclear power plants (Chernobyl) where the consequences of failures cover vast regions and go beyond the boundaries of individual countries, sometimes extending into the oceans, a common property resource. As we shall see later in this chapter, international treaties are one mechanism used to solve many of these common property and transboundary risk and hazard issues.

Prevention *versus* response

Another complicating factor for hazards management is the issue of prevention or response. In the US millions of dollars are spent annually for hazard prevention. For example, the majority of households have an array of insurance (health, life, car, fire, liability) to prevent economic loss from risks and hazards. Safety considerations govern every aspect of American life in particular, from the food we eat and water we drink, to where we work, and how we get there. Many argue that American society is over-regulated, sanitized, and obsessive about safety. Is it better to prepare in advance or wait until the risks pose serious threats to the health and well-being of people and the environment? This question raises one of the more fundamental issues in hazards management. Should management options be designed to protect people and the environment before a risk event occurs and to lessen the impact of the risk should it occur, be examples of prevention strategies, or simply respond after the accident or disaster?

When and where one intervenes in the hazards-management system—before the event takes place (prevention) or afterward (response)—is a key issue in hazards management. Central to the debate is a discussion over long- *versus* short-term goals and needs since each influence the variety of available management alternatives. There are many opportunities for intervening in the management system. Kasperson, Kates and Hohenemser (1985, p.62), for example, identify six: modify technology, prevent initiating events, prevent outcomes, prevent exposure, prevent consequences, mitigate consequences. A good example of how this operates is illustrated by a case study of automobile safety (Fig. 4.2).

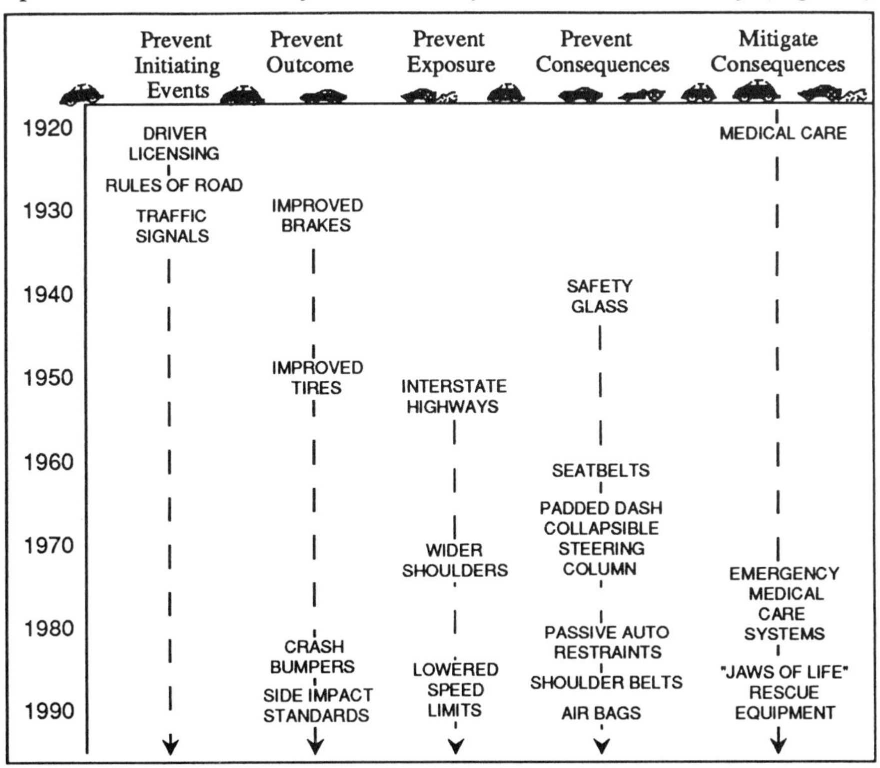

Fig. 4.2 Hazard control for automobile safety. Safety innovations are routinely incorporated into hazard-management systems in an effort to reduce injuries. Modified from Kasperson, Kates and Hohenemser 1985, p. 62.

As new safety innovations are introduced such as the reduction in speed limits on interstates (ostensibly to conserve fuel, but it had a side benefit of reducing traffic fatalities), air bags, or 'jaws of life' rescue equipment, they quickly become incorporated into the hazards-management system as both prevention and response elements.

Crisis management is the pre-eminent mechanism for managing most technological hazards. Invariably, each accident or incident is handled separately as a discrete event, necessitating the appropriate emergency response. This crisis-management approach may lead to unsound decision-making that is burdensome, costly, later regretted, or sometimes reversed. The Bhopal accident and the implementation of SARA Title III legislation in the US is a good example.

The Emergency Planning and Community Right-to-Know Act of 1986 (SARA Title III) was passed in aftermath of the 1984 chemical accident at Bhopal, India. Concerned with the possibility that a similar accident could happen in the US, this law had far-reaching provisions ranging from mandated emergency planning, to accident notification, emissions reporting, and of course, right-to-know provisions (Table 4.3).

Unfortunately, the implementation of SARA Title III has proven overly ambitious and costly. Emergency plans at the state and local levels were supposed to be in place by autumn 1988, but began appearing on a piece-meal basis during the autumn of 1991. The reluctance of the planning effort by local officials is largely due to lack of funding to undertake the endeavour. Another aspect of the law was the requirement of an annual assessment of toxic emissions. The Toxic Release Inventory (TRI) was first released by EPA in the autumn 1989 and covered the 1987 reporting period (USEPA 1989a). Not all facilities reported however, skewing the data by region and chemical, resulting in overestimates in others. Subsequent reports have reduced some of these problems (USEPA 1990a, 1991b). The TRI emissions reporting is one of the more successful elements of the programme.

From an industry perspective, SARA Title III has been a nightmare. The mere reporting of chemicals on hand meant that the industry would soon be regulated where it had not been previously. Industry was also concerned as the data became public, they would be increasingly liable should an accident occur. What used to be an 'in-house problem' and the cost of doing business, now became a public issue. Industry was also concerned over proprietary information in their reporting of the type and quantity of chemicals on-site and emitted. Rather than giving away trade secrets, many producers listed the maximum quantities that were available, preferring stricter regulatory control over an exposé of their processes and products.

While the basic reporting requirements of SARA Title III are necessary and have provided valuable information to the EPA and researchers, it has come at a cost to industry, who must hire consultants and other

Table 4.3 Main elements in Title III: The Emergency Planning and Community Right-to-Know Act of 1986.

Emergency Planning
- Defines local emergency-planning districts.
- Establishes local emergency planning committees (LEPCs) and mandates broad-based membership in them.
- Requires development of emergency response plan by autumn 1988.
- Provides technical assistance for planning efforts.

Notification
- Requires immediate notification of a hazardous substance release included on the EPA's list and/or subject to reporting under CERCLA.
- Specifies method of initial notification and information needs (chemical, acute toxicity, quantity, time, duration, health risks, etc.).
- Requires follow-up written notice with more specific information on response actions, anticipated health risks, and updates of verbal reports.

Community Right-to-Know
- Requires facilities to maintain material safety data sheets (MSDS) and submit these to LEPCs, fire departments and state planning commission.
- Requires facilities to submit emergency and hazardous chemical inventory form to LEPC providing estimates of quantity of all chemicals on site and general locations.
- Allows LEPCs to request additional information from facilities on specific substances.

Emissions and Release Reporting
- Requires EPA to develop an inventory of toxic chemical releases with 10 or more full-time employees in industries with Standard Industrial Classification codes 20–39.
- Requires EPA to establish a toxic chemical release form with name, location, type of business, category of chemical use, quantities present, waste treatment, quantity entering the environment annually, and certification of the accuracy of the report.
- Requires EPA to maintain a national inventory of toxic chemical releases.

Source: Cutter 1987, p. 15.

risk-professionals to insure they are in full compliance with the law. Whether or not we are more prepared for accidental releases than we were a few years ago is debatable. Since SARA Title III was drafted in response to public outrage over Bhopal, it failed to consider transportation-related accidents and the full range of toxic chemicals. Time will tell if this crisis-oriented approached to toxics management will be successful or not.

Another complicating factor of crisis management is the media focus on the acute event (Three Mile Island, Bhopal, Tylenol poisonings, airline disasters). The heightened awareness of the public to individual disasters, diverts attention away from the more chronic, pervasive, and everyday risks (air pollution, acid rain, chemical contamination, car safety). Finally, while catastrophes are the most effective means to stimulate hazard and risk control, all too often, repeated crises are needed to spurn governments and industry into action. While the seemingly random event can be ignored or justified as a rare occurrence, if the same type of failure occurs repeatedly, it is less easily ignored.

Equity

Perhaps no other issue has stymied hazard managers more than the concept of equity, be it social, regional, or economic. There is an uneven distribution of risks within society resulting in uneven exposures. Some people and places are simply more at risk by virtue of their location, occupation, or lifestyle. Moreover, hazards place disproportionate burdens on individuals and communities. Restitution for accidental exposures or environmental degradation follows the same pattern. Why are certain risks and hazards tolerated in the workplace but not elsewhere? The need for social justice and hazards management that reduces inequities in risk exposure, risk reduction, and risk compensation form the basis for new thinking about technological hazards (National Academy of Engineering 1986). The fairness question is equally applicable to hazards management. Fairness and equity form the basis for many of the contemporary management controversies over high-level nuclear-waste repositories, siting of hazardous-waste landfills, municipal solid-waste incinerators, and greenhouse gas reductions. Consideration of equity is one of the key emerging issues in hazards-management policies at levels from local to global.

Techniques

There are a wide range of hazards-management techniques. I will briefly describe just a few of these under broad generic headings (e.g. legislation, regulation or rules, adjudication or law suits, economic, planning, insurance, and dispute resolution) to illustrate their approach.

Legislation

Legislation is the most time-honoured method for managing risks at both the local and national levels. At the international scale, multilateral treaties, directives, and compacts accomplish the same thing as local or national laws, although their enforcement is more by coercion rather than legally binding. In the 1960s, less than 2% of all the laws passed by the US Congress were hazard-related. During the next two decades, slightly more were passed (3 and 4% respectively) (Table 4.4). As the public

Table 4.4 Technological hazards legislation in the US.

Congress	Year	Total laws	No. hazards laws	% Hazards laws
91	1969–1970	695	19	2.7
92	1971–1972	607	12	2.0
93	1973–1974	649	21	3.2
94	1975–1976	588	12	2.0
95	1977–1978	633	35	5.5
96	1979–1980	613	26	4.2
97	1981–1982	473	25	5.3
98	1983–1984	623	26	4.2
99	1985–1986	664	18	2.7
100	1987–1988	713	38	5.3
101	1989–1990	650	24	3.7
102	1991–	83	6*	7.2

* Five of the laws were appropriations for the Gulf War, the remaining law was a proclamation celebrating Earth Day 1991.
Source: Johnson 1985; Congressional Quarterly Service 1979–1990; Congressional Quarterly Service 1991.

becomes more sensitized to risk and hazards, the legislative response slowly catches up.

The most important risk-management statutes in the US are listed in Table 4.5 and govern a wide range of hazards. At the international level, the situation is more complex since multilateral treaties rarely have an enforcement element, a feature common to national law. While treaties are binding on signatory nations, there is no supreme enforcement body other than the World Court. Even then, the findings of the World Court are not binding on the countries involved if they choose to ignore them. Despite these caveats, there have been a number of international treaties concerned with managing technological risks and hazards (Table 4.6).

Regulation

Legislation, particularly in the US, provides a broad-based legal policy that requires subsequent implementation. The establishment of administrative agencies, rules and regulations, and permitting under legal statutes provides the regulatory approach to technological hazards (Portney 1990). These regulations are often mandated by specific laws, or instigated by regulatory agencies under their statutory authority or legal mandate. In either event, they often carry civil and criminal penalties for violations and are a formidable tool in managing technological risks and hazards.

Monitoring for compliance with rules and laws is one task of the regulating agency. The other task for regulators is enforcement. Enforcement

Table 4.5 Selected US risk and hazards laws.

Date*	Law	Primary risk/hazard provision
1966	Federal Hazardous Substances Act	Established the Consumer Product Safety Commission (CPSC) which regulates toxic consumer products
1970	Poison Prevention Packaging Act	Mandates child-proof packaging
1970	Occupational Safety and Health Act	Established OSHA, regulates workplace hazards, toxic chemicals in the workplace
1972	Food, Drug, and Cosmetics Act	Safety of food, drugs, cosmetics, colour additives, medical devices
	Section 346 (a)	Pesticide-residue tolerances for human food and animal feed
1972	Consumer Product Safety Act	Protects against dangerous consumer products
1972	Ports and Waterways Safety Act	Regulates waterborne shipment of toxic materials
1972	Maritime Protection, Research, and Sanctuaries Act	Prohibits ocean dumping
1976	Lead-based Paint Poison Prevention Act	Restricts use of lead-based paint in federally funded housing
1976	Toxic Substances Control Act (TSCA)	Effects of chemicals on the environment, regulates chemical risks, reviews all new chemicals
1980	Comprehensive Environmental Response, Compensation, and Liability Act (CERCLA)	Established Superfund for cleaning up hazardous-waste sites
1982/1987	Nuclear Waste Policy Act	Site selection for high-level radioactive waste repository, narrowed choices to Nevada
1984	Resource Conservation and Recovery Act (RCRA)	Identifies and characterizes hazardous wastes, generation, and transport, includes a cradle-to-grave tracking system
1986	Safe Drinking Water Act (SDWA)	Standards for 83 chemical contaminants

1986	Superfund Amendments and Reauthorization Act (SARA)	Title 3 emergency planning for air-toxic releases
1987	Clean Water Act (CWA) (Water Quality Act)	Toxics in water, regulates non-point sources of chemical contaminants
1988	Federal Railroad Safety Act	Safety and handling of hazardous materials by rail shipment
1988	Federal Insecticide, Fungicide, and Rodenticide Act (FIFRA)	Pesticide regulation, registration, labelling
1990	Clean Air Act (CAA)	Reduces hazardous air pollutants by 90%, sets emissions standards for 189 air toxics
1990	Asbestos School Hazard Abatement Act (AHERA)	Monitors asbestos removal and abatement in schools and other public buildings
1990	Oil Pollution, Prevention, Response, Liability and Compensation Act	Sets liability limits, mandates double hulls, international protocols for spill response
1990	Hazardous Materials Transportation Act	Rules for safety, placards, and handling of hazardous materials during shipment
1990	Pollution Prevention Act	Establishes a waste-management hierarchy to reduce waste at the source

* Date first passed or most recent reauthorization.
Source: Shapiro 1990; Congressional Research Service 1979–1990.

Table 4.6 Selected international treaties for hazards reduction and control.

1963 Moscow: Nuclear Test Ban (treaty banning nuclear weapon tests in the atmosphere, in outer space, and under water).
Prohibits atmospheric, underwater nuclear weapons tests and other nuclear explosions; prohibits tests in other environmental media if radioactive debris would be present outside of the country's territorial boundaries. 122 signatories, exception: France.

1972 London: Biological and Toxin Weapons (convention on the prohibition of the development, production, and stockpiling of bacteriological (biological) and toxin weapons and on their destruction).
Prohibits acquisition and retention of biological agents and toxins not justified for peaceful purposes, prohibits delivery means for such agents for hostile purposes or during armed conflict. 123 signatories, exception: Israel.

1972 London: Ocean Dumping (convention on the prevention of marine pollution by dumping of wastes and other matter).
Controls ocean dumping in a marine environment in waters seaward of the inner boundary of the territorial sea, prohibits certain materials ('black list') from being dumped (radioactive, biological toxins, mercury, cadmium, oils, plastics), regulates ocean disposal of others, establishes mechanisms for liability assessment and dispute resolution. 72 signatories, exceptions: Iran, Iraq, Indonesia.

1977 Geneva: Environmental Modification Convention (prohibition of military use of environmental modification techniques).
Prohibits signatory nations from engaging in military or other hostile actions against the environment or using environmental-modification techniques in warfare that have widespread and long-lasting effects. 50 signatories, exception: Iraq.

1978 London. MARPOL (Ship Pollution) (protocol of 1978 relating to the International Convention for the Prevention of Pollution from Ships, 1973).
Modifies 1972 London Convention by reducing pollution from ships including oil, chemicals, and plastics. 55 signatories, exceptions: Iran, Iraq, New Zealand.

1979 Geneva: Transboundary Air Pollution (convention on long-range transboundary air pollution and its related protocols).
Adoption of national policies for air pollution especially transboundary pollutants. 32 signatories.

1984: EMEP Protocol (protocol to the 1979 convention on long-range transboundary air pollution on long-term financing or the cooperative programme for monitoring and evaluation of the long-range transmission of air pollutants in Europe [EMEP].
Coordinates data collection and assessment of acidic deposition in Europe. 32 signatories.

1985 Helsinki: SO_2 Protocol (protocol to the 1979 convention on long-range transboundary air pollution on the reduction of sulphur emissions or their transboundary fluxes by at least 30%).
Signatory nations agree to reduce sulphur dioxide emissions by 30% of 1980 values by 1993. 21 signatories, exceptions: US, UK.

1987 Sofia: NO_x Protocol (protocol to the 1979 convention on long-range transboundary air pollution concerning the control of emissions of nitrogen oxides or their transboundary fluxes).
Binds signatory nations after 1994 to 1987 emissions levels of NO_x, also requires further action to reduce emissions by 1996. 27 signatories.

1980 Vienna: Convention on the Physical Protection of Nuclear Material.
Ensures the safe transfer of nuclear material and provides standardized measures for the physical protection of nuclear materials. 28 signatories, exceptions: Israel, France, UK, India.

1981 Geneva: Convention Concerning Occupational Safety and Health and the Working Environment.

Prevents accidental and health injury by monitoring the causes of hazards in the workplace. 12 signatories.

1982: World Charter for Nature.
Military activities damaging to the environment are to be avoided, advocates environmental protection from war and other hostile activities, advocates measures to prevent toxic discharges or radioactive-waste releases into the environment. 111 signatories, exception: US.

1985 Vienna: Ozone Layer (Vienna Convention for the Protection of the Ozone Layer).
Acknowledges the air as a resource requiring protection, provides for research on ozone-layer modification and its affects on human and biological health, monitors ozone depletion, and develop control activities to reduce depletion. 68 signatories, exceptions: Israel, India.

1985 Geneva: Convention Concerning Occupational Health Services.
Calls for establishing and maintaining a safe working environment, adaptation of the work environment to the capacity of workers according to their physical and mental health. 11 signatories.

1985 Rarotonga: South Pacific Nuclear Free Zone Treaty.
Establishes a nuclear-free zone in the region, free from radioactive waste pollution, weapons testing, and transport of nuclear materials. 10 signatories, exception: US.

1986 Geneva: Convention Concerning Safety in the Use of Asbestos.
Prevents and controls asbestos exposure of workers and protects them against health hazards due to occupational exposures. 10 signatories, exceptions: US, UK.

1986 Vienna: Convention on Early Notification of a Nuclear Accident.
Nations agree to provide relevant information about nuclear accidents as soon as possible to minimize transboundary consequences. 76 signatories, exceptions: Pakistan, Libya.

1986 Vienna: Convention on Assistance in the Case of a Nuclear Accident or Radiological Emergency.
Provides prompt assistance to affected countries to minimize health and environmental damage. 76 signatories.

1987 Montreal: Protocol (CFC Control) (protocol on substances that deplete the ozone layer).
Requires a 50% reduction in CFC production by 1999, allows for small increases in CFC consumption in the USSR and developing countries. 69 signatories, exceptions: China, India.

1990 London: Adjustments to the Montreal Protocol.
Strengthens CFC and halon controls, mandates 50% reductions by 1995 of CFC and halons, total phase-out by 2000.

1990 London: Amendment to Montreal Protocol.
Extends controls to cover new ozone-depleting substances, provides financial

mechanisms for implementing provisions of the treaty, provides a timetable for reductions in carbon tetrachloride and methyl chloroform emissions.

1989 Basel: Convention on Hazardous Wastes Movement (convention on the control of transboundary movements of hazardous wastes and their disposal). Restricts and controls international traffic in hazardous waste, export countries need written consent from import countries and countries traversed, export country manage shipment in environmentally sound manner, countries can prohibit imports of hazardous waste. 52 signatories, exceptions: Australia, Belgium.

1991 Bamako: Convention on the Ban of the Import into Africa and the Control of Transboundary Movements of Hazardous Waste within Africa.
Bans the import of hazardous waste into Africa and regulates transboundary movements of hazardous waste already in Africa. 17 signatories, exceptions: South Africa, Morocco.

1992 Rio de Janeiro: Climate Change Treaty.
Limits emissions of greenhouse gases, hopefully to 1990 levels, industrial countries are required to help finance technology transfer to developing countries to reduce greenhouse emissions, nations are required to formulate and update national mitigation programmes. 150 signatories.

Sources: UNEP 1989, 1991; OECD 1991a, 1991b; Rummel-Bulska and Osato 1991; World Resources Institute 1992; Stevens 1992.

action takes a number of forms. First, cases can be referred to other agencies (in the US this is the Department of Justice) for criminal prosecutions. For example, between 1982–1988, the EPA referred a total of 258 cases to the Department of Justice. During this same time period, 125 were prosecuted. In those cases where defendants were charged, the conviction rate was 72% (Russell 1990). Another mechanism is administrative action or civil penalties. In 1991 alone, 3416 actions were initiated, the majority for violations of the Clean Water and Safe Drinking Water Acts (Table 4.7) (Council on Environmental Quality 1992).

We have already seen a number of problems with the regulation of human health and safety, especially pesticides. Safety has a high priority within many regulatory agencies including the US EPA and the Occupational Safety and Health Administration (OSHA). However, EPA and OSHA must develop and regulate an increasingly large list of chemicals and hazardous substances; the list is now well over 500 substances. They are bound to miss a few.

One example is metam sodium. On 14 July 1991 a spill of metam sodium (an herbicide) occurred as the herbicide was being transported by rail. The train derailed and the tankcar plunged down a ravine near Dunsmuir, California (Ainsworth 1991; Ainsworth and Lepokoswki 1991). The tankcar was punctured and 19,500 gallons of the herbicide entered the Sacramento River. The US Department of Transportation which regulates

Table 4.7 Administrative enforcement actions by environmental law.

Year	CAA	CWA/ SDWA	RCRA	CERCLA	FIFRA	TSCA	EPCRA	Total
1972	0	0	–	–	860	–	–	860
1973	0	0	–	–	1274	–	–	1274
1974	0	0	–	–	1387	–	–	1387
1975	0	738	–	–	1614	–	–	2352
1976	210	915	0	–	2488	0	–	3613
1977	297	1128	0	–	1219	0	–	2644
1978	129	730	0	–	762	1	–	1622
1979	404	506	0	–	253	22	–	1185
1980	86	569	0	0	176	70	–	901
1981	112	562	159	0	154	120	–	1107
1982	21	329	237	0	176	101	–	864
1983	41	781	436	0	296	294	–	1848
1984	141	1644	554	137	272	376	–	3124
1985	122	1031	327	160	236	733	–	2609
1986	143	990	235	139	338	781	0	2626
1987	191	1214	243	135	360	1051	0	3194
1988	224	1345	309	224	376	607	0	3085
1989	336	2146	453	220	443	538	0	4136
1990	249	1780	366	270	402	531	206	3804
1991	137	1745	364	269	300	422	179	3416

CAA = Clean Air Act, first passed 1970; CWA/SDWA = Clean Water Act first passed 1972 and Safe Drinking Water Act first passed in 1974; RCRA = Resource Conservation and Recovery Act first passed in 1976; CERCLA = Comprehensive Environmental Response, Compensation, and Liability Act (Superfund) first passed in 1980; FIFRA = Federal Insecticide, Fungicide, and Rodenticide Act first passed in 1947; TSCA = Toxic Substances Control Act first passed in 1976; EPCRA = Emergency Planning and Community Right-to-Know Act (Title III, Superfund Amendments and Reauthorization Act of 1986), first passed in 1986.
Source: Council on Environmental Quality 1992.

interstate transport, does not classify metam sodium as a hazardous material, so no extra precautions were necessary. Yet, if this material is spilled in large quantities into water, it becomes lethal to aquatic life. In the Dunsmuir spill, more than 100,000 sport fish were killed, and aquatic life along a 45-mile stretch of river was heavily damaged. At one point, the spill threatened local water supplies. In the aftermath of the disaster, both the National Transportation Safety Board, the Chemical Manufacturers Association (an industry trade organization) and the US EPA are discussing ways to improve rail transportation and more tightly regulate and expand the definition of hazardous substances to include those that are potential water pollution hazards such as metam sodium. Adoption of the recently ratified MARPOL convention that identifies 500 chemicals as pollutants necessitating special marking and handling is one such effort. Improved design of tankcars, and more stringent safety regulations is also forthcoming in response to the spill (Morris 1991a; Hanson 1991).

Internationally, many countries have their own versions of environmental protection agencies, with regulatory control. In Canada, for example, it is Environment Canada. In Venezuela it is the Ministry of Environment and Renewable Natural Resources (MARNR) (White 1991). In the United Kingdom, a pollution control agency, known as Her Majesty's Inspectorate of Pollution (HMIP) was created in 1987. The purpose is to provide a comprehensive approach to pollution control and abatement under one directorate (O'Riordan 1988).

Adjudication

In managing hazards, the judicial system is often called in to mediate disputes. Litigation has become yet another tool for managing risk. With the advent of 'toxic torts' (an adaptation of common law that allows people to claim injuries to their health, property, or the environment from industrial activities) individuals can seek monetary damages from those industrial firms responsible for their injuries. In this civil action, the injured party (plaintiff) attempts to prove that one or more industrial firms (defendants) are responsible for the injury either by negligent conduct or failure to act. In either instance, they are liable for damages. In addition to compensatory damages, plaintiffs can also sue for punitive damages if they can prove the defendant's activities involved intent, malice, gross negligence, or fraud. The transformation from public to private risks and the increasing importance of toxic torts in American law has been based on improvements in hazard identification and risk assessment making it much easier to establish cause and effect (Huber 1986).

Courts are often asked to resolve claims against industry by local residents or the federal government. Liability for damages, over the Bhopal incident, for example, will be determined in both US and Indian courts. Negotiated settlements, such as those involving Exxon, the state of Alaska, and the federal government over the *Exxon Valdez* oil spill in Prince William Sound are increasing in frequency. Another example is a negotiated settlement between IMC Fertilizer and OSHA where the former paid US $10 million in fines (the original violations were US $11.95 million) in connection with a 1991 explosion at a facility in Sterlington, Louisiana (Kemezis 1991). This is the largest fine ever imposed by OSHA. Earlier record fines included a US $2.8 million against Union Carbide for wilful violations of safety in a 1990 explosion and fire at an ethylene oxide plant in Texas that killed one person and injured 32, and a US $4 million fine against Phillips Petroleum for a 1989 explosion at its Pasadena, Texas, facility (Kemezis 1991; Lepkowski 1991).

The impetus for negotiated settlements is to avoid delays in correcting the hazards. A more cynical view holds that companies are never found truly at fault, and thus emerge with a relatively clean slate. The fines are just the cost of doing business. Nevertheless, the companies are still liable for any civil suits filed by victims of the accidents, which in all likelihood would also be negotiated settlements.

Economic

Using the market to set fees is another management technique used for environmental hazards. Differential fee structures have long been used to manage hazardous waste and solid waste. Special fees for accepting out-of-state wastes that cover actual disposal costs and then some has been a long-accepted practice (see Chapter 6). Now however, the costs of transboundary disposal of hazardous waste is more expensive as many states and countries are now charging extra because of the environmental risks involved. In Alabama, for example, a US $72 per ton surcharge is placed on all out-of-state hazardous waste that is disposed of in the state. Since Alabama is an importing state, it feels it has the right to exact such a surcharge to compensate itself for the risks. Opponents are challenging this, and the case is now before the state Supreme Court. Similar problems exist at the international level with the shipment of hazardous wastes from industrialized to developing countries (see Chapter 6).

The use of private market and other economic strategies are fast becoming the preferred option for managing technological risks, especially in the US. Under the guise of polluter pays, pollution prevention, pollution reduction or risk reduction, the principle is the same. The costs of hazardous-waste disposal, pollution, as well as safety and adverse environmental impacts from accidents, has forced industry to modify its production processes to more effectively manage risks and hazards. It is now more cost-effective to reduce pollution at its source (the factory) than to dispose of the by-products (pollution into the air or water, subject to increasingly stringent environmental controls) and wastes. Similarly, it is also more cost-effective to develop risk-management protocols to enhance worker and environmental safety should an accident occur.

Other economic strategies aimed at reducing technological risks and hazards include consumer boycotts and consumer-driven demand for 'environmentally friendly' or risk-free products. The labelling of products according to their environmental impacts during the product's entire life cycle (initial manufacture to disposal) have gained popularity around the industrialized world (OECD 1991c). While the criteria are different, ecolabels are now in widespread use In Germany, for example, 3300 products are certified as 'environmentally correct' under the Blue Angel programme. In Canada the Environmental Choice labelling is now three years old, while in Japan 441 products are given the Eco-Mark label. In the US concern over the misapplication of the terms recycled, recyclable, biodegradable, and reusable as marketing tools led to increased calls for uniform labelling and certification programmes. One such effort, is the Green Seal programme designed to foster a national standard for environmentally friendly products (Hayes 1990). Another is the Green Cross programme (Salzhauer 1991).

Tax incentives are becoming increasingly popular hazards-management strategies. Enforcement, or command and control as it is often called, doesn't work very well and is always subject to delays as offending industries take their complaints to the courts. Offering tax incentives or tax abatements to industry to reduce emissions is now being tried in Louisiana. An environmental scorecard is developed for each facility, using such factors as past compliance with regulations, jobs to emissions ratios and so on. Bonus points are awarded for voluntary reductions in emissions. The scorecard is then used to determine what percentage of the tax abatement the facility will receive and would range from 50 to 100% (Morris 1991b). The programme has just been enacted, but there are already indications that significant decreases in emissions will accrue.

The power of the free market and capitalism in reducing technological risk and hazard is unparalled but it also has the same power to create risks. One startling example is the formation of Chetek in the former Soviet Union (Broad 1991; Potter 1991). With the rise of capitalism within the Commonwealth of Independent States, it was only a matter of time before entrepreneurs took over. One such scheme was the formation of Chetek, a private holding company endowed with the exclusive rights to peaceful nuclear explosives. Derived from the Soviet nuclear weapons design centre, the Chetek Corporation (named after the Russian words for man, technology and capital), is offering to sell underground nuclear blasts as a means for hazardous- and radioactive-waste incineration. Prices would vary between US $300–$1200/kg and would occur about 1 km underground. Despite its technical merit, this proposal has many worried about the proliferation of nuclear material, equipment, and technical know how. It is, however, one of the easiest and cheapest means for disposing of much of the highly toxic hazardous waste.

Planning

Planning is a hazards-management technique that has two main applications. The first is to reduce the threat of the hazard through land-use controls or buffer zones. The second is to anticipate the threat and develop emergency-response plans.

Land-use planning is one of the most contentious enterprises in local American politics. Public opposition to a host of facilities be they solid-waste landfills, hazardous-waste facilities, incinerators, refineries, or airports has created a quagmire in the siting of these locally unwanted land uses (LULUs). The siting task is made even more complicated when risky technologies are involved such as nuclear power plants. In a review of land-use planning, LULUs, and risk analysis Popper (1983) suggests that in order for risk analysis to become more effective in the public sphere of decision-making it should focus on consequences of the LULU in question and develop strategies to compensate the neighbours of LULUs rather than rely on expected value criterion. The notion of host-community benefits is one such example. Here communities are compensated for housing a LULU such as a landfill. The compensation can take a variety of forms (cash payments, tax rebates, public services improvements). Rather than opposing the siting of these LULUs, many communities might actually compete for them (see Chapter 7). While theoretically a good notion, the issue of equity poses some problems for this planning approach, since more often than not, it will be the economically depressed and/or socially disadvantaged communities that would seek such restitution.

Another form of land-use planning is the creation of safety or buffer zones around hazardous facilities. This buffer-zone concept is already used in the United States around commercial nuclear power plants where development is prohibited immediately adjacent to the facility (within a 2-mile radius). The buffer-zone idea has been recently tried in the United States near Plaquemine, Louisiana, where Monsanto recently purchased a neighbouring community and relocated its residents. The creation of toxic parks, akin to national parks, is one extrapolation of this concept (Cutter 1992). A 2-mile buffer zone around a hydrofluoric acid plant in Matamoros was recently ordered by the Mexican court. The Quimica Fluor facility has nearly 10,000 families living around the plant, and has had one fatal explosion in 1980, but none since (Heller 1991). In Bhopal, India, part of the reason for the significant loss of life was the encroachment of squatter settlements adjacent to the facility, an endemic problem in many developing countries.

Emergency-response planning is another tool to reduce risks from technological failures. After the 1979 TMI accident, for example, revisions to emergency-response plans were mandated for all facilities, and the planning area enlarged to a 10-mile zone, with a secondary ingestion pathway zone delineated for a 50-mile radius. The SARA Title III programme mandates emergency planning for chemical releases. Even industry has been coerced into emergency planning efforts through its Chemical Awareness and Emergency Response (CAER) programme. This programme enhances facility-local community cooperation in the development of emergency plans. Community involvement is further promoted by industry trade groups in their Responsible Care initiative, largely a public relations promotion on how concerned the industry is in community safety and right-to-know (Heller 1992). In Europe, the 1982 Seveso Directive and subsequent revisions tackles emergency preparedness and response to major technological hazards accidents. Emergency plans for handling and transporting hazardous materials and land-use controls are also under consideration (Chynoweth 1991).

There is an extensive social science literature on emergency planning, warning systems, and effectiveness of emergency planning that is too voluminous to review here. The best overviews can be found in (Church and Norton 1981; Cutter 1984; Lathrop 1981; Perry, Lindell and Greene 1981; Sorenson, Vogt and Mileti 1987; Foster 1980; May and Williams 1986; Comfort 1988; Sorenson and Rogers 1988; Sorenson *et al* 1988; Drabek and Hoetmer 1991). Broader practical guides and governmental manuals are found in DePol and Cheremisinoff 1984; Fawcett and Woods 1982; US DOT 1984; USNRC 1980; USFEMA 1981; and USEPA 1985.

Insurance

Loss sharing (or insurance) has long been practised in natural hazards management (e.g. flood insurance, earthquake insurance), and in personal property and safety (health insurance, car insurance, homeowners insurance). As a mechanism for distributing losses over a wide segment of the public, insurance is extremely effective. However, in the case of technological hazards, the liability for damages is much greater and responsible parties can, in fact, be identified. As a result, injured parties can recoup their losses by seeking compensation from the 'responsible' party or industry. This is much more difficult in the case of natural hazards when the events themselves are 'acts of God'.

In court cases punitive damages often exceed compensatory ones by orders of magnitude. Toxic torts, thus prove to be an economic liability to most industries. As a result, insurance is fast becoming one of most necessary techniques for managing technological hazards (Kleindorfer and Kunreuther 1987; Britton and Oliver 1991). Limiting their corporate liability is the goal for many industries. Costs have increased for the corporate insurance buyers as American society becomes increasingly litigious. Municipal costs have escalated as well, so much so that many communities refuse to open local swimming pools because of increased costs of liability insurance. Liability claims for negligence are most often heard in state courts. Redress from liability claims based on environmental contamination, land-use regulation, or violations of federal constitutional rights are more frequently than not mediated by federal and state courts (Pine 1991).

Dispute resolution

Right-to-know laws have had an enormous effect on informing the public about technological risks and hazards. As the public becomes more informed, industry is increasingly on the defensive. To thaw public opinion, many chemical companies are taking a more proactive stance and voluntarily reducing emissions. For example, Hoechst Celanese has recently announced a goal of reducing toxic chemical emissions at its 21 US production facilities by 70% over the next five years, at an estimated cost of US $500–600 million (Sternberg 1991). The goal is to enhance their image as an industry leader in emissions reductions and to demonstrate to the public its commitment to environmental protection. Anticipating mandatory emissions requirements that may be forthcoming, Hoechst Celanese is attempting to garner as many public-relations benefits as possible.

Environmental mediation is one facet of dispute resolution where aggrieved parties attempt to settle differences outside the judicial or regulatory system. The command and control mentality of many regulatory agencies is far less successful than in the past, prompting new avenues for discourse and resolution of environmental conflicts. Adversarial proceedings are being replaced by negotiated rule-making where industry, the regulatory agency, and environmentalists meet to achieve a consensus on the promulgation of specific regulations. In many respects, this is a win-win situation. By having involvement from the beginning, industry and environmentalists (normally occupying polar positions on the issue) can insure that their concerns and needs are met from the outset. The result is increasing acceptance of the regulations and a lessening of costly review procedures and legal challenges. The use of negotiated regulations or 'reg-negs' will increase as both industry and environmentalists find them the most efficient means for reducing technological hazards and risks.

Summary

The management of technological risks and hazards is a complex process fraught with many unresolved issues such as equity, transboundary considerations, scientific uncertainty, basic management approaches, and the proverbial bottom line of industry. As a result, the tools or techniques used in managing these hazards are as highly varied as the hazards themselves. They can range from very simple procedures such as doing without the technology or substituting a less risky technology, to very complicated schemes, involving regulation, monitoring, enforcement, or restitution. As we have seen in the preceding chapters, there is no simple solution to coping with technological hazards.

We know that geography matters and that social acceptability and management alternatives will vary depending on spatial scale. The social construction of risk, its contested nature, and the complexities, constraints, and opportunities provided in understanding technological hazards are important. In the remaining chapters I will provide individual case studies that illustrate these themes by looking at the stories of chemical hazards (Chapter 5), hazardous waste (Chapter 6), and nuclear technology (Chapter 7).

5

Winds of death

> *'But the tragedy near Milan was sudden, direct, concentrated, and awesome. It symbolizes the era of the chemical plague just as surely as Hiroshima signaled the tragedy of the atomic age. ... We could leave Seveso physically, but we could not leave it emotionally.... Somehow I felt Rachel Carson should have seen this place. But perhaps, as she wrote in* Silent Spring, *she already had'* (Fuller 1977, Forward, pp. 161–162).

First, there was Seveso—a runaway chain reaction that released a toxic plume of dioxin on the Italian countryside in 1977. More than 500 cases of burns and chloracne were reported, but no immediate fatalities. The release, lasting 2–3 minutes, permanently contaminated an area of 17 km^2 (Marshall 1987). While certainly a watershed event, Seveso is not the only recent instance of a large-scale chemical accident (Brown 1987).

The 1984 methyl isocyanate release from a Union Carbide pesticide-manufacturing plant in Bhopal, India, epitomizes some of the impacts of technological hazards. More than 2700 people were killed immediately, while another 100,000 or more were injured, some permanently. The interaction of this highly complex manufacturing system and semi-skilled operators, a lax regulatory environment, and local land use that permitted dense settlement near the plant, set the context for this tragedy. These factors should have been seen as precursors to the likely failure of the system. Surprisingly, none of the parties involved (plant designers, operators, corporate management, or public officials) anticipated an accident of this magnitude, despite many warning events (leaky valves, corroded pipes, staff failures in following procedures, maintenance shutdowns) that happened repeatedly.

Accidental releases occur daily, some small and unnoticed, others resulting in significant damage to both people and their property. The 1992 hexane explosion in Guadalajara, Mexico, that killed 180 and injured 600 more is just one of the latest examples. Hexane and possibly gasoline were dumped into the sewers by local industry (even PEMEX, the state oil company, is implicated). This set the stage for the blast that affected a mile-long area in one of Guadalajara's poorer areas, the Reforma district, parts of which were completely destroyed (Golden 1992).

Furthermore, these accidental releases are not new but have a rather long history that coincides with the development and diversification of the chemical industry itself. Despite one of the better occupational safety records in US manufacturing, the chemical industry has taken the brunt of the public's chemophobia; a conflict arising from differences in public perception and scientific views.

This chapter describes the evolution of chemical hazards which are the products of changes in the chemical industry and how society uses chemicals. The industrial and economic contexts coupled with the social and political milieu (regulation, acceptance of risks) help us to understand where chemical hazards occur and what people and places are most vulnerable to them. Before we can understand the impacts and management of these chemical hazards, however, we must first understand how they are created. To do so, we need to examine the historical and spatial evolution of the chemical industry to illustrate the geographic distribution of the hazard, how this has changed over time, and what places and people are most vulnerable to these hazards and why? In the last part of the chapter, I will focus on one type of chemical hazard—toxic clouds—to illustrate some of the constraints in identifying and managing these risks and hazards.

The production of chemical hazards

There are many types of hazards that are associated with the chemical industry but they generally fall into two categories: incidents/accidents and waste products. The initiating events for accidents/incidents are quite varied and range from rapid-onset events (1–30 seconds) such as bleves (boiling liquid expanding-vapour explosions) and fires and fireballs, to events with slightly longer onset times (minutes to hours) such as toxic clouds and fumes. All result in acute exposures. Longer developing (months to years) hazards such as hazardous-waste generation and disposal, and low-level toxic releases result in chronic population exposures. The severity and magnitude of the hazards are a function of the type of technological failures (burst pipe, high-pressure explosion), and the location (fixed-site such as an industry, or transportation related), both of which affect the spatial extent of the hazard. Chemical accidents or incidents can occur anywhere where chemicals are manufactured (industrial plants), transported (rail, truck, ship), stored (warehouses), or used (local swimming pools). This versatility in use makes their management extremely difficult. More than any other technological hazards, chemical hazards require a thorough understanding of the scale and context (social, historical, political, economic, environmental) of chemical use because of the source of the hazard is so randomly distributed.

Technological change or innovation is one of the primary driving forces in our socialization from a dispersed agrarian population to an urbanized industrial one. Since the Industrial Revolution, the pace of technological innovation has accelerated resulting in profound changes in social relations, physical landscapes, and society–nature interactions. The spatial incongruence and worldwide impacts of technology are now much greater than during the Industrial Revolution. As the rate of technological change increases and the technology itself becomes more complex, there is a concomitant increase in unintended and unanticipated consequences (Headrick 1991). More sophisticated and often more dangerous industrial processes produce both immediate and longer-term risks to human and ecological health, and more recently planetary health. Clearly the benefits of industrial growth, outweighed the human and environmental health risks of industrial failures. Nowhere is this more evident than in the chemical industry.

There is a distinct periodicity to the development and growth of the chemical industry, first in Europe and later in the US. This growth helps to explain the increased frequency of chemical accidents, one type of hazard produced by the chemical industry. Prior to 1900, the worldwide chemical industry was dominated by the British soda works and German dyestuffs sectors. After 1900 the industry became much more diversified and experienced rapid growth. This growth was due to a number of factors: (1) technological innovations and improvements in science; (2) the internationalization of the industry and the development of multinational corporations; (3) mechanization that allowed for increased efficiency in production; and (4) the abandonment of batch operations in favour of continuous-flow operations that required changes in design, purity of feedstocks, and pressure to temperature ratios (Haber 1971).

Product life cycles also played an important role in the early transformation of the industry as well as its spatial evolution (Markusen 1985; Chapman 1991, 1992; Auty 1984). A product life cycle has four stages (Fig. 5.1). Stage 1 (research and development) is when the innovation/product is conceived, tested, patented, and commercialized for production. Spatially, the industry is concentrated in only a few locations reflecting the distribution of inventors, innovations, or raw materials. In Stage 2 (new product stage), the industry experiences rapid sales as the new product is introduced to the market where the innovating firm clearly monopolizes sales. The core location is reinforced and enhanced as skilled labour moves to the area, and subcontracting relationships develop. An example of this is the early development of the computer industry in Silicon Valley, California, where the innovation (silicon chip) was first developed and manufactured, creating a 'high-tech' agglomeration of economic activity in the region.

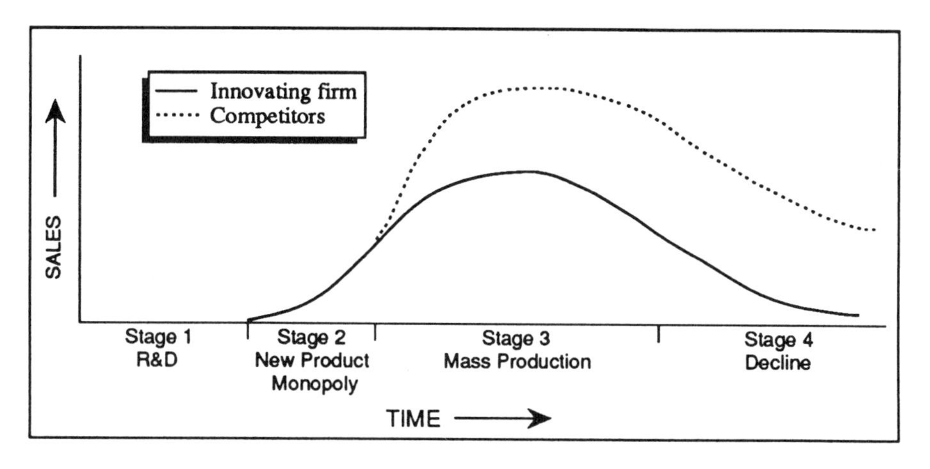

Fig. 5.1 Product life cycle. The product life cycle phases (research and development to product saturation) are a useful analogy to understand the evolution of the chemical industry.

Stage 3 (maturity) occurs when competitors enter the market and the innovative firm's sales level off. As the technology becomes standardized and the product is mass produced, there is a need to reduce costs, thus the industry becomes more dispersed, seeking locations with sources of cheaper labour in order to offset declining sales. In the final stage of the product life cycle, the product declines and eventually disappears, setting the stage for a new product/innovation. Often the manufacturing process becomes obsolete and companies are unwilling to invest in any future innovations or build more efficient machinery. Spatially, the firm reduces capacity, and may continue to sell the product to both domestic and foreign markets, but at a very low level.

Clearly, the product life cycle is not the only factor that helps explain the spatial evolution of the chemical industry. Governmental intervention is also a factor, especially in countries that do not rely primarily on free-market responses. A good example is the chemical industry in the Commonwealth of Independent States (CIS). The former Soviet Union has abundant hydrocarbon resources, providing it with a comparative economic advantage. Yet the inability of the centrally planned economy to respond quickly and concentrate investment resources to keep pace with technolgical innovations has severely limited its competitiveness in global markets (Sagers and Shabad 1990). The CIS chemical industry is concentrated in relatively few locations (central Russian region around Moscow, Donets-Dneiper and Ural region. Volga region, Belarus, western Ukraine) due to raw materials and energy, water, labour, and transport costs. Despite rapid developments in the post-war years, the location of the chemical industry continues to concentrate in existing industrial centres and/or metropolitan regions (Fig. 5.2). Agglomeration economies and

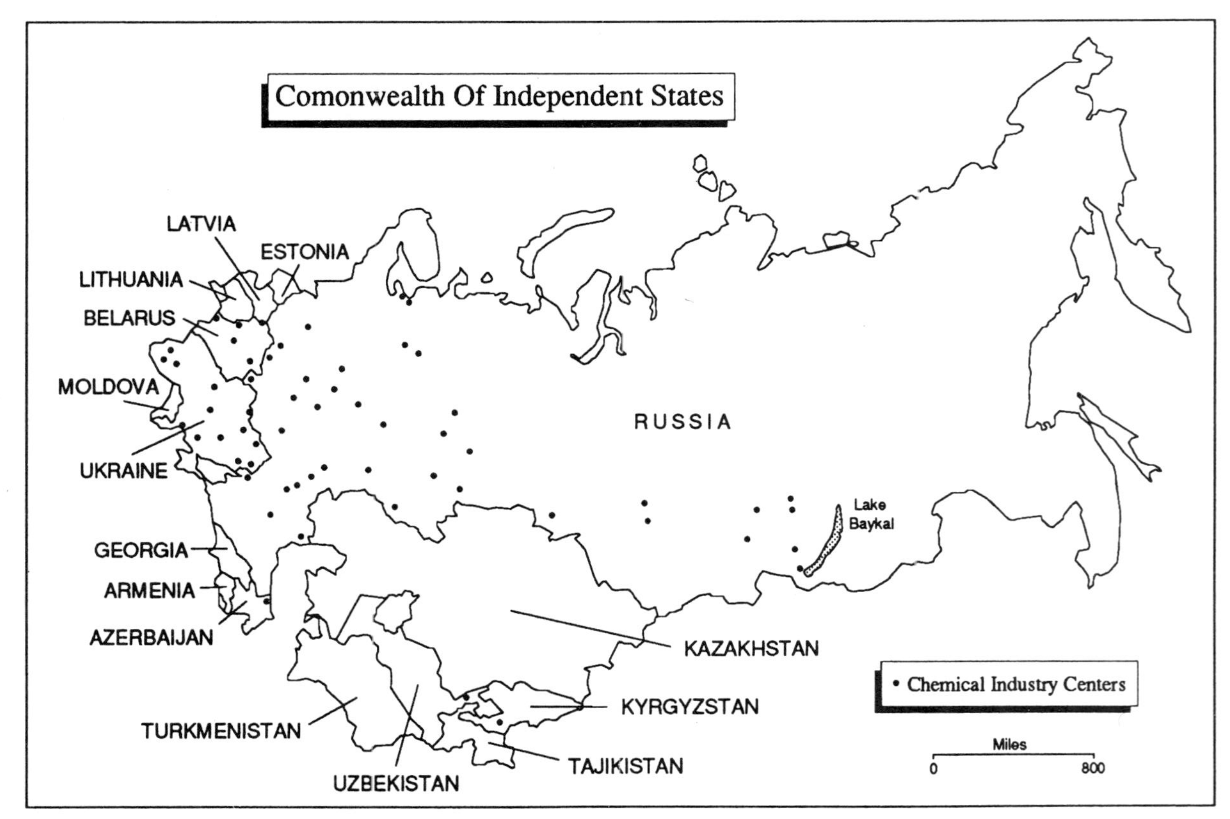

Fig. 5.2 Centres of chemical production in the former Soviet Union. The largest concentration of the chemical industry is in Russia and Belarus. Four republics (Russia, Ukraine, Belarus and Kazakhstan) account for more than 90% of the CIS chemical output. The largest chemical complex is the Mogilev Chemical Fibre plant in Belarus, employing more than 20,000 workers. Source: Sagers and Shabad 1990; Mitchell 1992.

the availability of advanced technical and engineering inputs have little influence on the spatial distribution of the CIS chemical industry (Sagers and Shabad 1990). This is not the case with free-market economies.

Historical and spatial evolution of the chemical industry

International developments pre-1900

Sulphuric acid was the first chemical produced on a commercial scale in 8th century Arabia, but it wasn't until the Industrial Revolution that the chemical industry came into its own, especially in Europe. In the early 18th century heavy organics, specifically alkali and soda ash, were the primary focus of the industry. Alkali and soda ash were in great demand in both the hard soap and glass manufacturing industries. The supplies of alkali and soda ash were from vegetable sources, and as demand far outstripped supply, innovations occurred. One of these, the LeBlanc process, used sulphuric acid on salt to make a synthetic alkali. While initially patented in France, the LeBlanc process was used extensively throughout Great Britain. The process was inherently hazardous, however, creating such by-products as hydrochloric gas (vented through the smokestack) and tank waste (which we would now call hazardous waste) that was dumped into the nearest water course. For every ton of alkali produced by the LeBlanc process, 1.4 tons of tank waste were discharged (Davis 1984). By 1863, the venting of hydrochloric acid and the dumping of tank waste were creating such serious health problems in surrounding communities that the UK Alkali Act was passed requiring a 95% absorption of gas by the factory. No sooner was the act passed when the industry realized that hydrochloric acid could be recovered to make bleaching powder, thus an unwanted by-product stimulated the development of an entire new subindustry (Davis 1984).

Another innovation that helped the soda industry was the development of the Solvay method for producing synthetic alkali. First produced in 1865 in Brussels, this method quickly became the preferred way to produce alkali because of less tank waste, cheaper raw materials and labour. The Solvay method, however, was more capital intensive and thus supplanted the LeBlanc process in most countries with the exception of Great Britain. The United Kingdom's inability to respond to this innovation eventually led to the demise of the soda works industry there.

There are other sectors of the chemical industry that were equally important prior to 1900. The dyestuffs sector developed by the Germans (Bayer, Hoechst, BASF [Badische Anilin und Soda Fabrik]) and Swiss (Ciba [Gesellschaft fur Chemische Industrie Basel], Geigy) in the 1860s were the first to produce synthetic dyes (anilines) from coal tar (Aftalion 1991). By 1900, the Germans controlled more than 90% of the world dye market. Coal tar was originally used to protect the hulls of wooden ships, and the gas it produced was a by-product. Later, the gas was utilized for heating and lighting and the coal tar became the waste product.

The explosives sector began much earlier. There was no real centre of strength or monopoly like the soda works and dyestuffs, thus competition among producers was fierce. Eventually, many firms began a period of cooperation by creating an oligopoly, the Gunpowder Trust, which established predetermined market areas among the world's munitions producers. The wireless phone and ocean-liner travel helped to consolidate the industry into two commercial giants: du Pont founded in 1804 in Wilmington, Delaware, and Noble Industries (United Kingdom) founded in 1871 in Glasgow. Pierre du Pont set up a powder-making plant near Wilmington on the Brandywine Creek, thus beginning his commercial enterprise. The hazards in production were enormous: saltpeter, sulphur, and charcoal were crushed separately and then combined. The powder was dried, sifted, compacted, and then packed into kegs. This mixture was highly explosive and flammable and any spark could set off an explosion; in fact there were many. Despite these hazards, the firm was enormously successful and certainly helped by world events that necessitated the company's product: Mexican War of 1848; Crimean War in Europe (du Pont sold to both sides); and the US Civil War (du Pont sold to the North only). Until 1900 du Pont held the monopoly for the American government's procurement of ammunition.

A new innovation by Alfred Noble significantly changed the munitions sector. Noble sought a safer way to detonate the gunpowder rather than with the highly unstable blasting agents such as nitroglycerin and gun-cotton. Noble's inventions (detonator in 1866, dynamite in 1867) revolutionized the munitions industry allowing for major social changes to occur because of the safety of dynamite as a blasting agent. Without these inventions, the building of the trans-Canadian railway, tunnels through the Alps, and the Panama Canal would never have been completed at the time. Although never quite able to crack the du Pont hold on munitions in the US, Noble Industries became the leading munitions manufacturer in the world.

Lastly, the pharmaceutical sector's early growth began in 1898 with the invention of aspirin by Bayer (Germany). The pesticide and fertilizer sectors were also in their infancy at this time. In the mid-19th century, for example, sulphur and a lime mixture were applied to thwart mildew, and arsenic compounds for eradicating the Colorado potato beetle were used (Davis 1984).

The 20th century and the American chemical industry

Pre-World War I: rapid innovations

In the US, the chemical industry was still in its infancy and was dominated by foreign capital and foreign technology prior to 1900. The country was a source of raw materials and a market for finished goods rather than the producer and exporter of goods and technology (Thackery and Bowden 1989). There were more than 8000 manufacturing plants employing around 191,000 people. Nearly one-quarter of these were small facilities employing fewer than 20 people. The value added by manufacturing in 1904 was US $282 million according to US Census figures (Table 5.1). While most of the companies were small, family-owned operations, there were a number of large, foreign-owned plants. Solvay, for example, operated plants in Syracuse and Detroit, while electrolytic alkali plants were found in Rumsford (Maine), Niagara Falls, and Wyandotte (Michigan). By 1900, there were distinct concentrations of chemical producers centred close to local markets (Philadelphia, northern New Jersey, New England), sources of cheap energy (Niagara Falls), or close to natural resources such as salt (Detroit), coal (West Virginia), oil (western Pennsylvania, Pittsburgh), or phosphate rock (South Carolina, Georgia, Florida). Surprisingly, there were few chemical producers in the Midwest and in the Gulf Coast states.

The pre-war years were characterized by rapid technological developments in the field. Herbert H. Dow's extraction of bromine from the brines in Midland, Michigan, set the stage for direct competition with the

Table 5.1 Changes in the chemical industry.

Year	No. operating manufacturing establishments*	No. employees (×1000)	Employees per plant (mean)	Value added by manufacturing (US $ million)
1904	8370	191.2	22.8	286.2
1909	10 380	235.4	22.6	401.4
1914	10 698	269.2	25.2	456.9
1921	8208	278.9	34.0	833.7
1929	9327	381.6	40.9	1737.3
1933	7297	302.0	41.4	1120.6
1939	8839	na	na	1818.9
1947	10 019	626.4	62.5	5317.0
1954	11 075	733.9	66.3	9549.9
1958	11 372	698.3	61.4	12 308.0
1963	11 996	737.4	61.5	17 586.1
1967	11 799	841.4	71.3	23 550.1
1972	11 425	836.5	73.2	32 413.9
1977	na	880.2	na	56 720.5
1982	11 901	872.6	73.3	77 314.8
1987	12 039	814.0	67.6	120 777.6

* US Standard Industrial Classification code 28.
Source: US Bureau of the Census 1963, 1972, 1982, 1987.

German chlorine monopoly. The electrolytic process and the formation of chlorine allowed Dow to undercut the German's market in the US. Another innovator was John F. Queeny who established Monsanto in 1901 to produce the synthetic sweetener, saccharin. Many other discoveries and improvements in production processes helped the American chemical industry gain international respect and markets. These included the discovery of Bakelite plastic (a condensation product of phenol and formaldehyde), from which billiard balls and piano keys were made. And of course there was George Eastman, a leader in the chemical processing of cellulose film and co-founder of Eastman Kodak.

War and inventiveness

It is quite obvious that armed conflicts stimulate technological innovations in many industries, and this is especially true in the chemical sector. War shortages in Germany, for example, led to a search for synthetic sources of nitrogen (for explosives and fertilizers) since natural sources were now restricted by trade embargoes. The Haber-Bosch process (direct synthesis of ammonia using nitrogen and hydrogen at high temperatures and pressures) was first developed in 1909 by a chemist at BASF with the assistance from an engineering colleague, Bosch (Haber 1971). The production of synthetic ammonia and nitrogen began at BASF's Oppau facility in 1915. Haber was also responsible for the development of chlorine gas, a new chemical weapon that killed 15,000 Allied troops during the 1915 battle at Ypres. The chlorine industry developed initially to manufacture the weapon but at the end of the war new applications were found for liquid chlorine (not gaseous) in consumer products such as bleach. German chemists were also noted for the development of mustard gas (1917), and chloropicrin (another toxic gas) first used in 1916 by the Russians, but later used as an insect fumigant. The munitions industry led by du Pont and Noble continued to flourish during this period as well.

In the US, World War I intervened to cut the supply of German organic chemicals. This created the demand for acetone, a chemical used to manufacture explosives. Carlton Ellis in 1916 discovered a method for using petroleum to make isopropyl alcohol (a raw material used to produce acetone) thus signalling the beginning of the petrochemical sector (Stobaugh 1988).

By 1914 the American chemical industry was larger than Great Britain's but still smaller than their German counterparts. The industry was still diffuse, however, with quite a few small works (average size workforce per establishment was around 25), generating US $457 million in revenues (Table 5.1). A large concentration of manufacturing remained in the Northeast and Great Lakes states. Alabama and Georgia maintained their stronghold in phosphate and fertilizer production, while California continued to grow as a diversified manufacturer of chemicals (Fig. 5.3a). The fertilizer sector was the largest of all chemical producers followed

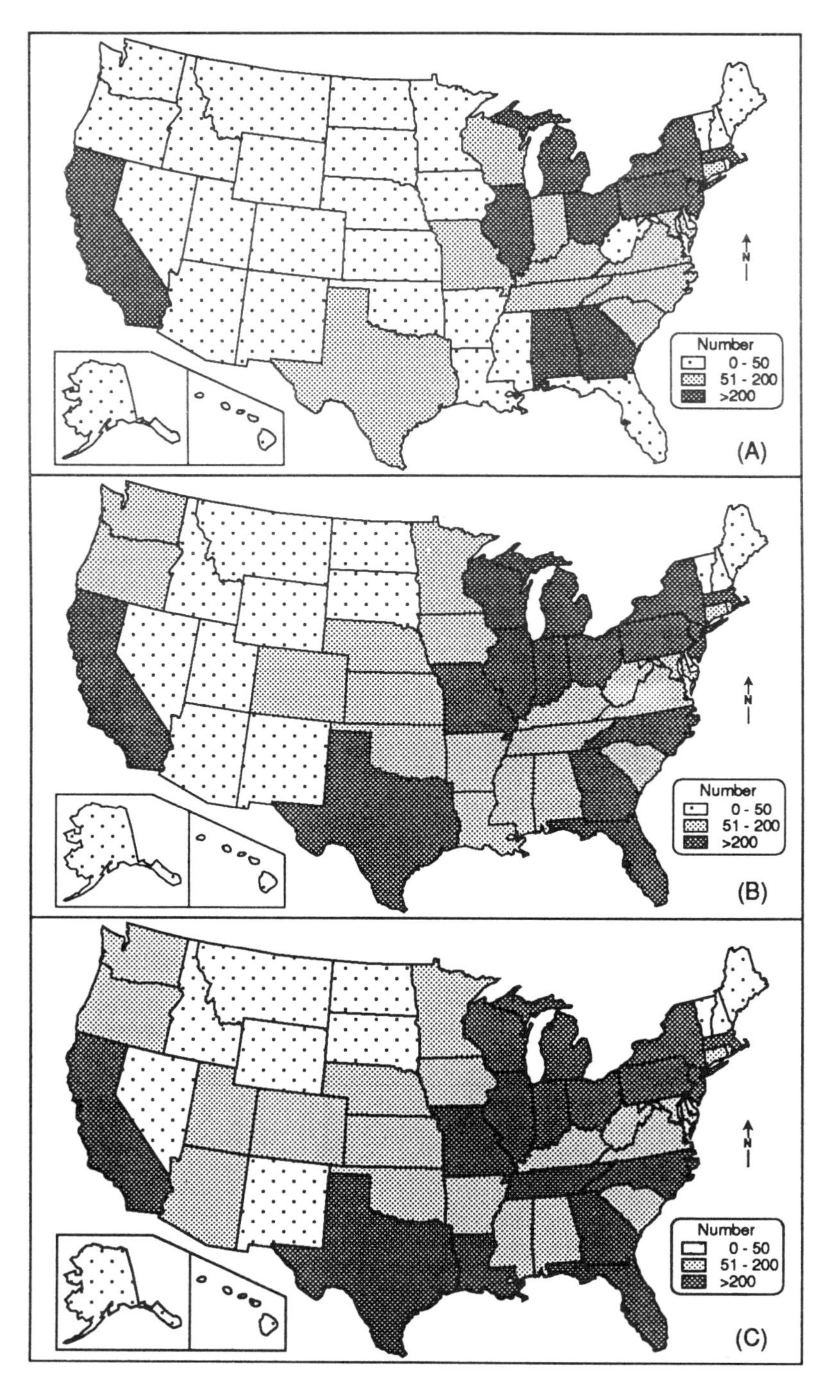

Fig. 5.3 US chemical manufacturing plants. The number of operating chemical manufacturing establishments for (a) 1914, (b) 1954, and (c) 1987 shows very little change in their distribution. Data from US Bureau of the Census 1965, 1972, 1987.

by the explosives sector. At the end of World War I, employment in the chemical industry had increased by 45% from the beginning of the decade, while revenues nearly tripled during the previous five-year period (Table 5.1). It was also during this period that the du Pont ammunitions monopoly was forced, through the Sherman anti-trust litigation, to create two new companies: Hercules Power and Atlas Power.

The interwar years: merger mania and diversification

The 1920s and 1930s were the decades of mergers and acquisitions as chemical companies grew and expanded their operations into new products and markets. This was true in Europe as well as the United States. Noble Industries, for example, merged with four of the largest British alkali firms to form Imperial Chemical Industries (ICI) in 1926. In Germany, mergers took place with Hoechst, BASF, Bayer, and AGFA merging to create IG Farben (Interessengemeinschaft Farbenindustrie). Within IG Farben, production was reorganized and specialized (all photo products for example were marketed under AGFA), and profits were used to diversify into other equally lucrative areas like synthetic fertilizers, plastics, and pharmaceuticals. Research took precedence over product manufacturing as a means for upgrading the company.

A similar pattern was followed in the US by du Pont who initially purchased General Motors; a ready-made market for its motor additives and auto finishes. Monsanto of St. Louis, began to diversify during this time as well into caffeine, vanillin, and aspirin production. The latter was extremely lucrative since Bayer's aspirin patent expired in 1917. Monsanto was also the first company to synthesize phenol (a main constituent of plastic) thus beginning their entry into the petrochemical field. Union Carbide and Carbon Corporation (an amalgam of three firms that came together in 1920) began to diversify into consumer goods, vanadium mills and mining, industrial gases, and acetylene plants. The result was that they cornered the market for ferro-alloys, electrodes, batteries, and flashlights. American Cyanamid Company, originally a fertilizer producer, also began to diversify into plastics, coal-tar dyes, catalysts, explosives, and pharmaceuticals.

Two important discoveries occurred during the interwar years that irrevocably changed the US chemical industry. The first was the development of polymerization, the process where small molecules or monomers are joined together to make larger ones (polymers) in long chains. This process enabled the development of synthetic rubber and more importantly, plastics. Synthetic polymers developed during this period included polystyrene (1930) and polyvinyl chloride (1931) both by German chemists, and nylon and polyethylene in 1938 by American chemists (Davis 1984). Since many of these polymers were derived from coal tar or cellulose bases, sources of new raw materials were needed in order to increase the mass production of these new polymers and their products.

The second major innovation was the process of catalytic cracking where crude oil and gas is decomposed by heat, pressure, and catalysts to create more feedstocks (raw materials used to create these synthetic polymers). This process results in a higher volume of products at lower costs. The fractions produced by natural gases and refining oil are made into the feedstocks as primary inputs into plastics manufacturing. For example, naphtha is a feedstock for solvents; ethylene is used for polyester fibres and plastic polyethylene; propylene is used for polyurethane foam; and benzene is used for fibres and styrene plastics. Catalytic cracking, by introducing new raw materials (petrochemicals) ultimately changed the nature of the chemical industry. The industry not only supplied chemicals used as agents in manufacturing, but they also began to make products that could be marketed directly to the consumer.

One offshoot of the synthetic revolution described above was the pesticide industry that already marketed directly to the consumer. The most significant innovation during the interwar period for this sector was the development of broad-spectrum pesticides. Concern over price fluctuations in species-specific pesticides such as calcium arsenate for boll weevils led to the notion of broad-spectrum pesticides with the ability to attack and kill many different species. In 1931, only 39 broad-spectrum pesticides were registered for use in the US (Davis 1984), while today there are thousands. DDT (dichloro-diphenyl-trichloroethane) was discovered by the Swiss firm, Geigy, in 1939 and was heralded as the ultimate broad-spectrum pesticide: efficient, long-lasting, and cheap. Broad-spectrum pesticides completely revolutionized the pesticide industry which marched full force into producing other synthetic organic agents such as the chlorinated hydrocarbons (DDT, dieldrin, chlordane) and the organic phosphates (parathion, malathion). These miracle pesticides were sold the world over with much fanfare and little concern (or knowledge at the time) about the human health or environmental impacts of their use.

The number of operating manufacturing firms remained fairly constant during these two decades, peaking in 1929 and bottoming-out two years later during the Great Depression (Table 5.1). The value added by manufacturing escalated from US $834 million in 1921 to more than US $1.8 billion by 1939.

Shortages and substitutions

As happened earlier in the century, World War II also stimulated technological developments in the chemical industry as a result of shortages in many products and the need to find substitutes. As the Japanese seized the Malaysian rubber plantations, both the Germans and the Americans experimented with synthetic rubber. In the US, the need for tyres for war vehicles helped to create the synthetic rubber industry almost from scratch. The petrochemical industry grew because of the US government's largesse in helping to finance much of the expansion (Stobaugh 1988).

The need for plastics, pharmaceuticals, and pesticides all stimulated growth in the chemical sector especially for Dow, Monsanto, and du Pont, the largest companies. The destruction of many of the German competitors during the war, the forced breakup of IG Farben (a condition imposed by the Allied forces), the development of the US oil and gas industry, and of course the affluent American market all contributed to make US petroleum companies world leaders after the war. By 1949, more than 10,000 manufacturing firms were operating, an increase of 13% from the previous decade. More people were employed than ever, and of course revenues increased (US $5.3 billion in 1947).

Unheralded growth and transition

There were very few technological changes in the industry during the 1950s and 1960s although a number of new synthetic polymers (Teflon in 1944, Dacron in 1949) were discovered. More important changes came as a result of the restructuring of the industry itself to meet the increased domestic demand and the search for new international markets.

A typical company prior to 1950 was family-founded and owned with mostly domestic plants, small headquarters staff, and no labour unions. In order to compete in the global market, many companies went public and established international divisions. Growth and diversification thus reshaped the industry from small enterprises that supplied processing chemicals or compounds to other manufactures (such as dyestuffs to the textiles industry, chlorine to the pulp and paper mills, pigments to paint makers, and fine chemicals for pill manufacturers) to much larger entities. Diversification allowed a company not only to produce the dye for the fabric, but the fabric itself (nylon, rayon); instead of the caustic for soap, it made the new synthetic detergent; or instead of supplying the processing chemicals for tanneries, the firm made the synthetic leather (Naugahyde). This transformation required newer and larger facilities and a constant source of raw materials. With this in mind, petroleum refiners entered the chemical business creating a vast array of potential markets for their petrochemicals (Spitz 1988). In the 1980s, the largest chemical producers were coincidentally oil companies.

Another innovation was the utilization of the mass media to market the chemicals. The selling of chemicals to American consumers became big business and advertising helped to blunt the potential impact of their use. Soap became synthetic detergent, later becoming surfactants; synthetic fibres became man(*sic*)-made fibres; and ethylene glycol became all-season anti-freeze (Jordan 1989).

It was clear during this period that economies of scale were driving the industry. The number of operating manufacturing plants increased slightly (6%) from the previous decade, while the workforce increased 14%. The average size of the plant increased from 66 employees to 71 employees by 1967. More than 35% of all manufacturing establishments employed 20 people or more compared to less than 25% two decades earlier. Equally important was the continued increase in revenues from US $9.5 billion in 1954 to US $23.6 billion in 1967. Regionally, the core of chemical manufacturing remained in the Northeast and Great Lakes states and California (Fig. 5.3b). However, fertilizer production expanded from Georgia into Florida. New petrochemical plants in Texas accounted for most of the growth during this period.

Another structural influence on the chemical industry during this period was the increased governmental regulation of the pharmaceutical sector as a result of the side effects of a number of drugs. The anti-diarrhoeal drug containing clioquinol used extensively in the 1950s particularly in Japan, resulted in an illness the Japanese called SMON. The side effects included a loss of motor abilities and blindness. However, the most notorious drug episode occurred in 1957 with the sedative, thalidomide. This over-the-counter drug was recommended to pregnant women and was extensively used in Europe and Japan. The drug produced severe foetal deformations in 8000–10,000 children. Although not widely used in the US (the Food and Drug Administration blocked the licence), the thalidomide tragedy led to stricter testing and regulation of all pharmaceuticals both domestically and abroad and led to tighter controls on product licensing and testing.

It was also during these decades that new chemical weapons were discovered and used. For example, Agent Orange (2,4,5–T and 2,4–D) the powerful defoliant containing dioxin was manufactured exclusively by Dow Chemical (1962–1972) for the US government for use in Vietnam. The unchecked growth of the industry was beginning to slow as the public became more fearful of chemicals. That fear was heightened in 1962 with the publication of Rachel Carson's *Silent Spring* which chronicled the dangers of persistent pesticides on the environment and the increasing resistance of pests to many of the compounds. This book in addition to the anti-war protests throughout the decade and consumer advocates like Ralph Nader, led to more public concern about chemicals and laid the groundwork for the environmental activism that followed during the 1970s.

Safety and environmental awakenings

Up until 1969, the chemical industry was not seriously challenged by the burgeoning public environmental awareness in the country. This all changed during the 1970s as a result of a number of highly publicized events.

First was the passage of the 1970 Occupational Safety and Health Act (PL 91–596). Prior to OSHA's passage, health and safety of workers was regulated by the individual states, with variations all over the country. Although known to pose serious health risks, lead, mercury, and asbestos were not controlled in the workplace until after the passage of this new law (Brinkman, Jasanoff and Ilgen 1985). The Occupational Safety and Health Administration was established and began standard-setting for workplaces with 10 or more employees. Physical (noise, temperature) as well as chemical agents were regulated. OSHA initially set standards on 400 chemicals based on prior work of the American Conference of Governmental Industrial Hygienists (Scott 1989). In addition, OSHA required workplace labelling of chemicals, safety training, recording keeping of accidents, and safety inspections. Finally, under the new law, a national institute of occupational safety and health (NIOSH) was established to provide health studies and new worker-exposure standards.

Second, a series of accidents resulting in occupational and environmental exposures heightened public concern over chemical use. In 1973 a truck delivered Firemaster (a fire retardant containing polybrominated biphenyls [PBBs]) manufactured by Velsicol Chemical Company to an agricultural feed plant in Michigan by mistake, instead of the dairy additive known as Nutrimaster. The bags simply got mixed up: similar names, same manufacturer. The result included thousands of dead farm animals throughout the state, deformities in many others, and widespread contamination that still lingers. It was also during this time, for example, that Kepone (a pesticide) contaminated much of the James River and poisoned workers in its Virginia manufacturing plant. In 1978 the EPA closed down a dibromochloropropane (DBCP) plant in Lathrop, California, after workers became sterile, and pesticide residues were found on surrounding crops and in the local groundwater. The production of hazardous waste and its indiscriminate disposal by the industry became apparent with the discovery of Kentucky's Valley of the Drums, and Love Canal, New York, increasing the public's chemophobia and providing renewed calls for tighter environmental regulations of the industry.

During the 1970s the chemical industry continued its growth and expansion, particularly into the petrochemicals sector. Despite the oil shocks of the decade, chemical producers were able to push the higher energy and materials costs on to consumers. Profits increased on the order of 16% per year from 1975–1980 (Chemical Week 1983). The profits were universally reinvested by companies into new plants (upgrading old ones, building new ones) with the predictable over-capacity by the end of the decade.

The 1970s were also a time for reorganization of the industry and decentralization. Prior to 1973 fixed costs were 80% of expenditures, while variable costs (such as feedstocks) were only 20%. After the oil crisis, these percentages reversed with variable costs increasing four-fold (Chemical Week 1983). As a result of the heavy reliance on foreign sources of oil, petrochemicals as a growth sector waned near the end of the decade as electronic chemicals, pharmaceuticals, and specialty chemicals became the new growth fields.

A number of examples serve to illustrate the reorganization and decentralization of the industry during this period. Du Pont generated

US $6 billion in revenues in 1973 largely based on its fibres (polyester, nylon) division that accounted for 35% of its total production. By 1983 du Pont generated US $33 billion; the majority of profits coming from the energy division as a result of du Pont's merger with Conoco. Monsanto began the decade with US $2.6 billion in profits (1973) with fibres and industrial chemicals their largest products divisions. By 1983 Monsanto's profits were US $6.3 billion primarily due to its agrichemical division with the success of its Roundup and Lasso pesticides, and its polymers division (Chemical Week 1983). At the same time, Monsanto moved much of its production (especially the agrichemical sector) overseas thus reducing costs and increasing profits.

By 1987 the chemical industry had expanded from the core area (Fig. 5.3c). Rapid growth occurred in Texas (74% increase in operating establishments since 1954) and Louisiana (37% increase) as new petrochemical facilities were constructed. New York with the largest number of operating establishments in 1954 with 1407, lost more than 47% of these by 1987. Other reductions in chemical manufacturing establishments were in Pennsylvania, Illinois, Iowa, and Nebraska. Regionally, increases in chemical manufacturing facilities occurred in the mountain west. Utah, Arizona, New Mexico, Idaho, and Montana all doubled the number of manufacturing establishments between 1954–1987. By 1987 the leading chemical manufacturing states based on number of plants were California (1423), New Jersey (912), and Texas (911). Revenues also continued to increase reaching US $77 billion in 1982, and US $121 billion five years later (Table 5.1). Du Pont, Dow Chemical, and Procter and Gamble were among the top ten world leaders in chemical production in 1989 garnering net profits of US $2.1 billion, US $2.4 billion, and US $1.2 billion respectively (Aftalion 1991). The leading producers include the three German giants Bayer, Hoechst, and BASF, and the British ICI, whose histories we described earlier.

Risk and hazards assessment

Accidents have always been a part of the industry. For example, the French freighter *Mont Blanc*, carrying 5000 tons of high explosives, was struck by another ship causing the explosion in Halifax, Nova Scotia's harbour in 1917. A 2 mile^2 area of Richmond (a suburb) was completely destroyed (Nash 1976; Morehouse and Subramanian 1986). In Oppau (Ludwigshafen), Germany, 560 people died and between 1500 and 1900 were injured when an ammonium nitrate explosion occurred at the BASF plant in 1921. Nearly three-quarters of the city was destroyed including 1000 homes, the plant itself, and warehouses (Marshall 1987; Morehouse and Subramanian 1986; Smets 1987). Twenty-two years later at the same site, a vapour-cloud explosion involving butadiene caused extensive damage within a 35,000 m^2 area, killing 57 and injuring 439 more. And five years later (1948) there was another gas (dimethyl ether) explosion that killed between 184 and 245 people, and injured between 4000 and 6000. Damage covered 40,000 m^2 (Lagadec 1982; Nash 1976; Marshall 1987).

One of the primary problems in assessing chemical hazards is the lack of data. There are a number of general overviews and compendiums that provide some historical information on technological failures and industrial crises (Nash 1976; Marshall 1987; Cashmann 1988; Lagadec 1982; Withers 1988; Smets 1987; Shrivastava 1987; OECD 1987, 1989, 1991a; UN Environmental Programme 1989, 1991; Davenport 1977). Most of these, however, focus on a wide range of industrial failures, rarely differentiating the term 'industrial'; sometimes it means chemical accidents and other times not. Also, the statistics rarely distinguish the medium of contaminate exposure (air, land, water, combination) and often involve restrictive time-frames (last five years). More importantly, only large accidents (those with high casualty numbers [fatalities or injuries], or those involving large evacuations) are normally reported. Thus there is a systematic underestimate of the actual frequency of chemical accidents worldwide.

There is no comprehensive data base of industrial accidents resulting in chemical hazards although there are a number of efforts currently underway. US legislation in the aftermath of Bhopal (SARA Title III and the Community Right-to-Know Act), as well as European Community's Seveso Directive, and the UK's Control of Industrial Major Accident Hazard (CIMAH) programme are developing national and regional data bases but are still in their formative stages. At the international level, the UN's International Programme on Chemical Safety (IPCS) helps in the development of national prevention programmes and assists in chemical emergencies. Technical assistance on chemical production, trade, and profiles of individual chemicals including toxic effects, controls, and environmental fates is provided by UNEP's International Register of Potentially Toxic Chemicals (IRPTC). More than 110 nations have participated in the registry, gathering and disseminating information (Cutter 1991).

Very little is known about the longer-term hazards such as waste production and disposal. We will examine this in the next chapter.

Airborne toxic releases

International patterns

As might be expected, there are conflicting reports on the trends in chemical accidents. In a review of major industrial crises (defined by 50 or more fatalities) Shrivastava (1987) found 28 major incidents, half of them occurring since 1979, with four happening in one year alone (1984). In a different review, the OECD (1991a) found a decline in accidents since 1975 claiming that better prevention and mitigation helped reduce accident frequency. The United Nations estimates claim about 200 serious chemical accidents yearly in OECD countries. These same estimates place worldwide totals during the last decade at more than 5000 deaths, 100,000 injuries and poisonings, and 620,000 evacuees from major chemical accidents. Shrivastava also claims that industrial crises are more common in developing nations because they lack the prerequisite industrial structure and safety mechanisms to cope with accidents. In a review of historical trends in chemical accidents, Cutter (1991) found a steady rise in accident frequency but a decline in severity (measured by deaths and injuries) in chemical accidents form 1900–1990. More than a third of the accidents took place in developing countries in Latin America and Asia. (Fig. 5.4).

The US hazardscape

The global pattern of accidents reveals little about the contextual nature of the hazard. Obviously, a more detailed analysis is warranted, yet there have been remarkably few attempts to examine the pattern of incidents at a more local level. Cutter and Solecki (1989), for example, examined airborne releases of acutely toxic chemicals at the state level using 1982–1986 data, while Cutter and Tiefenbacher (1991) provided an analysis of urban areas using the same data base. Sorenson and Rogers (1988) and Sorenson (1987) examined local preparedness for chemical accidents and evacuations from off-site releases of chemicals using 1980–1984 data. Finally, Solecki (1990) examined rural counties in four states using 1982–1989 data.

While useful, these studies fail to provide the requisite historical and industrial contexts within which to understand the increasing hazards of chemical use. Colton (1990a,b) examines relict industrial landscapes to reconstruct the patterns of industrialization, but fails to relate this to the pattern of accidents (or failures in the technology creating a hazard to surrounding populations).

A staggering number of chemical accidents were recorded during the 1980s with estimates ranging from 295 for the 1980–1984 period (Sorenson 1987) to more than 10,933 for the entire decade (Industrial Economics 1989). These figures include a much broader definition of accidents, not just those involving air releases.

To reduce the level of uncertainty regarding the number and location of incidents involving airborne releases, data on explosions and/or toxic clouds were derived from published sources such as those mentioned above and from a review of the *New York Times* index from 1900–1979. Since the focus of the analysis is on the airborne release of

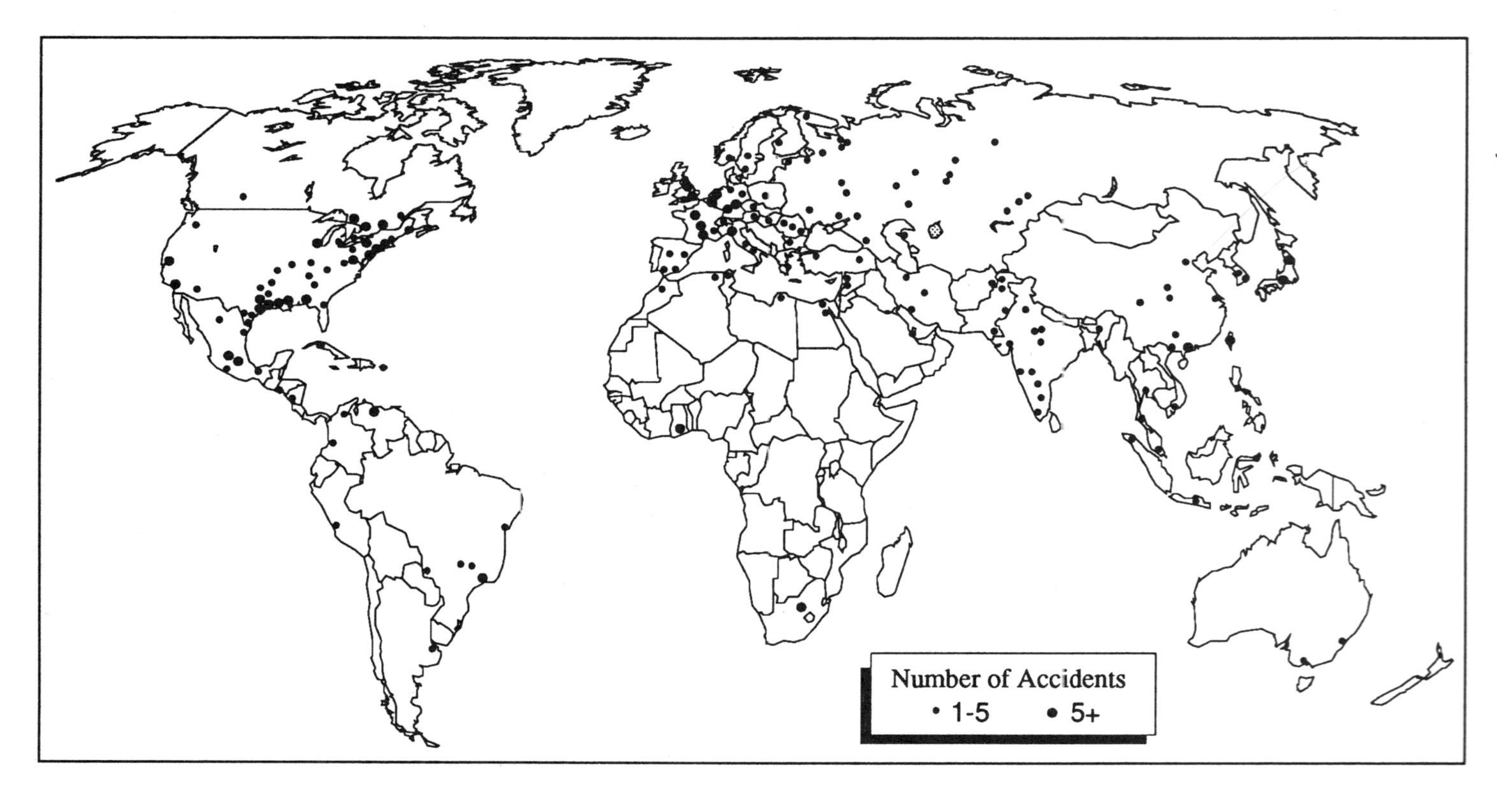

Fig. 5.4 Chemical accidents 1900–1990. About 50% of the accidents occurred in the US. An additional 20% were in developing countries in Latin America and Asia.

chemicals, any explosion was included as were incidents that involved the formation of a vapour or toxic cloud. Chemical spills of other types (water or land) were specifically excluded from consideration. These data are obviously biased, both in terms of their spatial coverage as well as their appearance in the media. Nevertheless, they provide a conservative estimate on the geographic patterns of these hazards.

Frequency and distribution

A total of 339 incidents were found during the period 1900–1990, compared to 333 worldwide during the same time-frame. The majority occurred in the 1970s and 1980s, accounting for more than 60% of the total (Table 5.2). Beginning in 1950, there is almost a doubling every decade in the number of chemical incidents. Again, this is not surprising, for most of the incidents originated at stationary sources such as manufacturing plants. Transportation (rail and truck) and pipelines had fewer numbers of accidents. There is a small decline in the total number of incidents during the 1980s.

Texas had the largest number of incidents (41), followed by California (30), New Jersey (28), Louisiana (24), New York (24), Pennsylvania (21) and Illinois (17). Geographically, those states with the greatest number of incidents are clustered in the Northeast (New York, Pennsylvania, New Jersey) and Gulf Coast (Texas and Louisiana), with outliers in Illinois and California (Fig. 5.4). Most of New England, the upper Great Plains, and the West are relatively accident-free.

Chemical types

More important perhaps than the number of total incidents, is the chemical involved. For our purposes, accidents were divided into six different classifications based on chemical type. Acute chemicals are those that cause immediate death or severe physical harm when people are exposed to them in vapour form. These include chemicals like ammonia and chlorine. In 1985, the Superfund Amendments Reauthorization Act (SARA) provided a listing of such chemicals that was used in our classification. Radiation was another category. Ammunition and explosives (including fireworks) was the third category, followed by oil and natural gas. The fifth category included a variety of chemicals that were identified but did not fit any of the previous four, and our sixth class were those agents that were unknown (Table 5.3).

The majority of the accidents involved acute chemicals from stationary sources during the 1970–1990 period. Regionally, Texas and Louisiana have the greatest frequency of acute chemical releases followed by New York. One interesting observation is West Virginia, which had only 11 incidents during the 90-year period, yet nine of these (or 82%) involved acutely toxic chemicals, obviously one indicator of higher risk.

Oil and natural gas explosions were the next most common form of

Table 5.2 Chemical accidents 1900–1990.

International

Year	Total	Acute releases*
1900–1909	2	0
1910–1919	19	0
1920–1929	18	3
1930–1939	20	2
1940–1949	25	4
1950–1959	27	8
1960–1969	36	14
1970–1979	89	41
1980–1989	97	38
Totals	333	110

United States

Year	Total	Stationary sources	Transportation	Pipeline	Unknown	Acute release
1900–1909	4	4	0	0	0	0
1910–1919	14	12	2	0	0	1
1920–1929	12	11	1	0	0	5
1930–1939	6	3	3	0	0	3
1940–1949	15	8	5	1	1	6
1950–1959	18	12	4	1	1	5
1960–1969	53	35	10	3	5	29
1970–1979	118	55	43	8	12	50
1980–1989	99	53	38	2	6	66
Totals	339	193	106	15	25	165

* Acute releases are vaporous chemical releases that cause immediate physical harm and possibly death to exposed populations. Acute-release chemicals follows SARA Title III list.
Source: Data compiled by author.

chemical incidents. The hazards here are not with exposure to toxic chemicals but the explosion itself. Again, the majority of explosions occurred in the 1970s. Those states with the highest frequency include Texas, New York, New Jersey, California, and Illinois.

Ammunition and explosives were more common in the earlier decades with the majority occurring between 1910 and 1919. Like oil and natural gas, the hazard is mainly from explosion not toxic exposures. Regionally, New Jersey, New York, Pennsylvania, and Maryland had the highest frequency of ammunition disasters.

Table 5.3 Types of chemicals involved and severity of accidents.

Chemical class	No. accidents US	No. accidents non-US	No. deaths (US)	No. injuries (US)
1. Acutely toxic	165	94	628	7746
ammonia	19	20		
chlorine	34	22		
ethylene	15	9		
misc. (SARA and CERCLA)	97	43		
2. Radiation	7	16	2	128
3. Ammunition and explosives	30	105	1088	2583
4. Oil and gas	89	73	551	2841
5. Other (identified)	39	37	223	1536
6. Unknown agent	9	8	95	56

Source: Data compiled by author.

Severity

The magnitude and severity of chemical incidents is often hard to gauge since estimates of death and injury, and property damage are of dubious reliability and fluctuate widely depending on the source of the statistic. In a review of different data on industrial disasters using a threshold of 50 fatalities, Shrivastava (1987) found 6936 deaths during the 1907–1984 period worldwide. His data did not include mine disasters, nor ammunition explosions.

In the US during roughly the same period, chemical incidents resulted in 2587 deaths and 14,910 injuries (Table 5.3). If we examine the fatality figures by chemical grouping a number of observations are in order. First, munitions explosions caused the most fatalities followed by acute releases. While less deadly, acute releases resulted in more injuries (7746) than any other class of chemicals. Radiation was relatively injury-free, causing only two fatalities and 128 injuries, although this is somewhat misleading since casualty figures only report on immediate deaths and injuries, not longer-term consequences of the exposure that result in death/injury years later.

The death and injury statistics are also misleading in the assessment of the severity of incidents. The largest disaster involving an acute chemical happened in Texas City, Texas, in 1947 with an explosion and fire on the ship *Grand Camp* that was carrying ammonium nitrate. The blast was felt 150 miles away, and the fire completely destroyed the city, including the Monsanto synthetic rubber factory that burned and released styrene fumes into the air. The number of fatalities ranged from 468 to 576 with between 1000 and 5000 injuries. If we exclude this one accident from our tally, we find that only 106 fatalities resulted from the other 167 accidents or less than one fatality per accident. In fact, only 25% of the acute chemical accidents resulted in any fatalities. Ammunition explosions, on the other hand, have a more consistent pattern of fatalities and injuries on a per accident basis. Most of these, like the Port Chicago, California, explosion in 1944, involved the loading or shipping of TNT for military use. One plausible explanation for the reduction in casualties is the increased use of precautionary evacuations. Until 1979 less than 20% of the incidents involved an evacuation. From 1980 onwards there were 62 incidents (63% of the total) that involved an evacuation, most of these from acutely toxic substances. More than 40% of the evacuees came from six incidents (Table 5.4).

Table 5.4 Largest US evacuations from chemical accidents.

Date	Place	No. evacuated	Chemical	Source
1986	Miamisburg, OH	40000	white phosphorus	T
1987	Salt Lake City, UT	30000	ammonia	F
1981	San Francisco, CA	30000	PCB-contaminated lubricating oil	P
1969	Tallahatchie, MS	30000	vinyl chloride	T
1980	Newark, NJ	26000	ethylene oxide	T
1988	Los Angeles, CA	23500	chlorine	F
1980	Somerville, MA	23000	phosphorus trichloride	T
1918	South Amboy, NJ	20000	TNT	F
1985	West Chester, PA	20000	pentaerythritol	F
1988	Springfield, MA	20000	chlorine	F
1982	Taft, LA	18500	acrolein	F
1987	Nanticoke, PA	17000	sulphuric acid	F
1988	Henderson, NV	17000	perchlorinated ammonia	F
1987	Pittsburgh, PA	16000	phosphorus oxychloride	T
1981	Blythe, CA	15000	nitric acid	T

Abbreviations for sources are F = fixed site such as chemical plant, industrial complex, warehouse; T = transportation by rail, truck, or ship; P = pipeline.

Oil and gas explosions fall somewhere between the two extremes mentioned above. While rather frequent events, the consequences (death and injuries) are greater than acute releases on a per accident basis, much less than ammunition explosions. The largest natural gas explosion occurred in 1944 in Cleveland. A liquid natural gas tank at the East Ohio Gas Company exploded due to structural weaknesses and ignited. The fireball destroyed 79 houses, two factories, 3600 cars and left 1500 people homeless (Marshall 1987; Nash 1976; Cashman 1988). As a result of fears about other explosions, liquid natural gas was eliminated for commercial use for nearly two decades.

Contexts of risk

It is clear that the frequency of chemical accidents is increasing nationally and globally, and that the toxicity of chemicals has increased as well, posing even greater dangers to surrounding communities. Having said this, it is important to provide a comparative perspective on all types of industrial accidents, not just chemical ones. According to government researchers, in the period 1976–1978 alone, more than 127,725 chemical-related illnesses and injuries were reported (Lowry and Lowry 1988). Most of these were in the manufacturing sector (all products) and involved explosions and fires. On an industry-by-industry basis, the chemical industry as a whole had one of the best safety records during this period (Lowry and Lowry 1988). Despite this, our data during the decade showed a total of 118 accidents, 43% involving acutely toxic chemicals that killed 32 and injured another 1885, and necessitated the evacuation of more than 50,000 people from ten of the accidents.

Industrial transformation and increasing toxicity

Earlier in the century, the chemical industry's production largely centred on heavy inorganics such as phosphate, and coal tars. The production of hazards was localized and in the form of waste products or occupational exposures to workers. As the industry diversified, we saw more special applications and the development of synthetics. There is a dramatic increase in the number of incidents after 1959 that coincides with the rapid development of the petrochemical and plastics sector during the 1950s and the increase in larger facilities with larger outputs and larger consumption of toxic feedstocks (Fig. 5.5). The frequency of acute toxic releases also rises significantly. The revamping of industrial processes, the construction of larger facilities, and the use of highly specialized and ultimately toxic feedstocks enhanced the production of hazards, particularly in the absence of any governmental control. This is true for airborne releases and explosions as well as for hazardous-waste production.

It is clear there is a relationship between the transformation of the chemical industry and the increased number of chemical disasters such as toxic clouds. The production of hazards accelerated along with the growth and diversification of the chemical industry. While accidents now make the news more frequently and our awareness of them is heightened, I argue that those hazards historically have increased in their relative frequency as well. There was a slight decline in the actual number of incidents from the 1970s through the 1980s, although there was a significant increase in the number of acutely toxic releases. More than two-thirds of all releases in the 1980s involved acutely toxic substances, whereas the percentage was much lower (42%) a decade earlier. Whether this pattern continues into the 1990s and beyond is open for speculation.

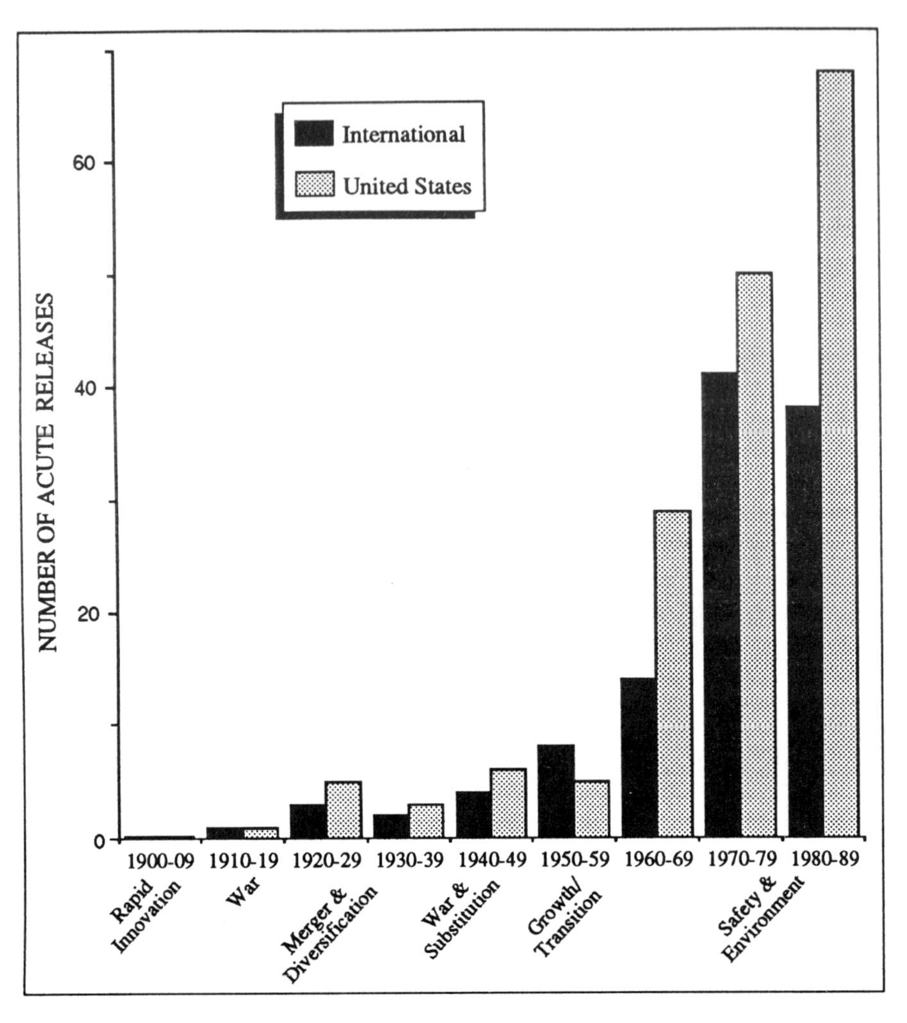

Fig. 5.5 Accidents involving acutely toxic substances. There has been a steady rise in the number of accidents involving these acute toxins that coincides with changes in the chemical industry (larger production units, more toxic feedstocks, diversification) as well as stricter health, safety, and environmental quality standards.

Vulnerability and scale

By virtue of their level of economic activity, some places are more likely to have chemical hazards than others; they are more vulnerable to them. In its broadest terms, vulnerability is defined as the likelihood that an individual or group will be exposed to and adversely affected by a hazard. The concepts of risk, preparedness and mitigation, and the social geography of affected populations are common elements found interwoven in many vulnerability studies.

Emergency preparedness can and often does reduce the risk of chemical hazards. Accidents are local and happen so quickly that there is often little time to call on extensive emergency-response networks. To facilitate local emergency-response, the US Congress in 1986 passed the Emergency Planning and Community Right-to-Know Act, SARA Title III (see Chapter 4). Even though federal funds were not appropriated for emergency planning, many communities now include chemical accidents as part of their overall disaster and hazards plans. Many of the obstacles to emergency planning for these events (lack of knowledge about chemical use in the community, lack of prior experience, unknown responses of residents) have been reduced through improved levels of risk communication to the public by industry and local emergency managers. These efforts include public education and frequent practice drills also involving the public.

There is no doubt that scale is an important consideration in assessing vulnerability. Utilizing the state scale to measure the degree of place vulnerability Cutter and Solecki (1989) identified two types of incidents (rural agricultural and urban industrial), but the spatial clustering of each was masked by generalizing to the entire state. Consequently, it was difficult to completely determine the underlying factors and processes that contributed to a state's vulnerability to these types of releases. Working at a larger scale, multi-county metropolitan areas, Cutter and Tiefenbacher (1991) differentiated high-hazard, medium-hazard, and low-hazard cities and began to identify some of the contextual factors, especially risk indicators that helped to explain the spatial distribution. However, even at this scale some inter- and intra-metropolitan comparisons were difficult.

Solecki (1990) examined county-level data for rural counties in four states (Alabama, Georgia, Iowa, and New York) between 1982 and 1989. He found a total of 262 incidents, mostly from stationary facilities. The majority of releases included acids, ammonia, or chlorine. Spatially, incidents were clustered in counties adjacent to metropolitan areas in two states (Iowa and New York) while there was a random distribution in the remaining two. Solecki also found a clear relationship between the number of incidents, the number of local facilities with hazardous materials, and greater levels of emergency response. His study was very useful in illustrating the relative importance of risk and preparedness in assessing vulnerability to these types of hazards.

Vulnerability is an elusive concept but we know that both form (places, people) and process (factors that enhance or retard it) are important. Clearly more conceptual development and detailed analyses of vulnerability to chemical hazards (specifically) or technological hazards generally is needed (see Chapter 8). We have seen that chemical hazards are embedded in contexts—political, economic, industrial, social, historical. In this regard, understanding and living with chemical risks requires an examination of the processes that create or contribute to their formation. Only then, can we begin to ascertain which places and people bear the burdens of hazards and are most vulnerable to them.

6

The global dumping ground

'Toxic nightmares have spread throughout the country, and the message is clear: chemical waste disposal too often is in the hands of people who aren't fit to deal with it' (Brown 1980a, p.120).

Environmental degradation by hazardous waste and chemical pollution is a by-product of our industrialized world. While the degree of contamination varies from place to place, the source is the same: the improper disposal of wastes from industrial processes and/or consumer products containing hazardous substances. In some places, the level of contamination is so great that the mere mention of the place—Love Canal, Times Beach—conjures up the image of a toxic wasteland (Brown 1980b). Abandoned waste sites are the current legacy of past industrial activities and their indiscriminate contamination. Chemical time bombs (the chemical build-up in soils and sediments without adverse impacts until a threshold level is reached or the ecosystem is somehow disturbed thereby releasing the chemical contamination) are another legacy (Stigliani *et al* 1991). Both are examples of intergenerational inequity (the activities of one generation severely impacting on the next). We are paying the environmental price for actions taken by our parents and grandparents; our children will bear the burdens of the technological hazards we create today.

There are geographic inequities as well. Hazardous-waste disposal is a major problem confronting the industrialized world. It has become a major business and regulatory headache as solutions are sought to reduce it. There is the ever-increasing trade in hazardous waste as states and countries become commodity experts in the global toxic export/import business, more often than not dumping it on the poor and powerless. The inequities in risk burdens and the tragic choices that confront society regarding hazardous-waste generation and disposal provide the focal point of this chapter. The inability to manage the hazards at the local level, or in the generator's own backyard, has resulted in a global game of hide and seek, often taking the form of toxic terrorism.

Affluence breeds effluence

No one knows how much hazardous waste is generated on a yearly basis, let alone where it goes: into the local air, land, sea, fresh water, or sent out of the state or country. While 2.1 billion tonnes of industrial waste are produced annually, only about 16% of it is hazardous (OECD 1991a) (Table 6.1). Geographically, most of the hazardous waste is produced in the United States. In 1987 238 million tonnes were generated in the US alone and more recent estimates suggest a range between 250 and 275 million tonnes annually (UNEP 1991). Even accounting for population size, the US is still the world's largest producer of hazardous waste, generating about 1 tonne per person per year (Table 6.1). Germany, Italy, Canada, and France round out the top five.

Table 6.1 Annual production of hazardous waste (billion tonnes annually).

Country	Population (millions)	Hazardous waste (1000 tonnes)	Tonnes per capita per year
US	249.2	238 327	0.96
Canada	26.5	3000	0.12
Japan	123.5	666	0.005
Australia	16.9	300	0.02
New Zealand	3.4	60	0.02
Austria	7.6	400	0.05
Belgium	9.8	915	0.09
Denmark	5.1	112	0.02
France	56.1	3000	0.05
Finland	5.0	230	0.05
Greece	10.0	423	0.04
Germany (west)	61.1	14 210	0.23
Ireland	3.7	20	0.005
Italy	57.1	3640	0.06
Luxembourg	–	742	–
Netherlands	14.9	1500	0.10
Norway	4.2	200	0.05
Portugal	10.3	165	0.02
Spain	39.2	1708	0.04
Sweden	8.4	500	0.06
Switzerland	6.6	400	0.06
Turkey	55.9	300	0.005
United Kingdom	57.2	2200	0.04
Eastern Europe	–	19 000	–
Rest of world	–	16 000	–
Total	–	308 318	–

Sources: Hazardous waste figures from OECD 1991a, b, pp. 139, 141. Population data from World Resources Institute 1992.

One of the major difficulties in arriving at these estimates is the lack of consensus on the definition of 'hazardous waste'. The other problem is accurate record keeping (where the waste is produced, how much, and where it ultimately goes). As a consequence, the estimates on worldwide trends in hazardous-waste generation vary widely. On a regional level, data are even more dubious as some countries keep better records than others, while in many countries there is no record keeping at all.

What is hazardous waste?

Hazardous waste is solid waste that has physical, chemical, or biological characteristics that cause or contribute to threats to human health (leading to serious illness or death), or adversely affect the environment when improperly managed (Goldman, Hylme and Johnson 1986). We typically think of hazardous waste originating from the chemical, mineral, and metal-processing industries. However, consumers such as you and I also generate hazardous waste—from the dry-cleaned clothes we wear, to the car repair shop we use, to the products we use around the house (Table 6.2) Our lawn care and gardening products like pesticides, oven cleaners, and other household cleaners all too often contain hazardous substances. While these products contribute a fraction of the total hazardous waste produced, it does add up, especially when these products are placed in local landfills rather than in specialized hazardous-waste facilities.

Table 6.2 Consumer sources of hazardous waste.

Products	Types of waste products
Plastics	Organic chlorine compounds
Pesticides	Organic chlorine compounds, organic phosphate compounds
Medicines	Organic solvents and residues, heavy metals (mercury, zinc)
Paints	Heavy metals, pigments, solvents
Oil, gas, and petroleum products	Oil, phenols, heavy metals, ammonia salts, acids, caustics
Metals	Heavy metals, fluorides, cyanides, solvents, pigments, plating salts, phenols
Leather	Heavy metals, solvents
Textiles	Heavy metals, dyes, solvents
Electrical equipment	Cyanides, heavy metals, caustics, solvents
Paper	Dyes, bleaches, solvents

Sources: USEPA 1980; Goldman *et al* 1986.

Unfortunately, not all substances that could be considered hazardous are regulated. At best, only a fraction of the toxic wastes that we generate are regulated under the myriad of pollution-control laws (see Chapter 4). The US EPA, for example, classify hazardous wastes based on four characteristics: ignitability, corrosivity, reactivity, and toxicity (Fig. 6.1). Radioactive waste is a special category and handled separately (see Chapter 7). Ignitability refers to substances that have flash points below 140°F and thus pose a fire hazard during management. Examples of these include many oils, oil-based products, solvents, and other highly flammable products.

Corrosivity means that the waste requires special containers such as steel drums since the waste could corrode standard ones. Corrosive substances also cause chemical burns to the skin and are especially dangerous to the eyes. Corrosive materials such as acid wastes need to be segregated as their potential to dissolve other waste is quite high. Examples of corrosive substances included hydrochloric acid, sodium hydroxide, and hydrofluoric acid. Reactivity refers to the explosive potential of waste. The material could react spontaneously, or interact with air or water and become unstable, resulting in a toxic gas, or an explosion. Examples of reactive substances include methane and chlorine. The final characteristic, toxicity, means that the substance poses a major threat to

Fig. 6.1 Characteristics of hazardous wastes. Four parameters are used to define hazardous wastes: ignitability, corrosivity, reactivity, and toxicity. Source: USEPA 1998, p. 85.

human health and the environment. Examples include arsenic and other heavy metals, and methyl isocyanate (a key ingredient in the insecticide, Sevin).

Depending on the particular law, the number of regulated substances and reportable quantities varies. Since there is no consistent definition of hazardous or toxic materials, regulations are often written on a chemical-by-chemical basis, or by groups of chemicals. The list of regulated substances, therefore, changes as more detailed health and environmental-impact data become available. For example, under the Toxic Release Inventory report mandated by the SARA Title III legislation, emissions of 302 chemicals in addition to 20 categories of chemicals are regulated. A total of 348 substances are regulated under the RCRA, while over 500 are covered under the Superfund legislation.

While the European Community provides a definition of hazardous waste, there are still individual country's interpretation of that definition. Estimates on the amount of hazardous-waste generation therefore range from 17 to 50% of all industrial waste. According to one estimate, the constituents of the waste include solvents (6–8%), waste paint (4–5%), waste with heavy metals (4–10%), acids (30–40%), waste oil (17–20%), and miscellaneous chemicals (17–39%) (OECD 1991b). Compounding the problems even further is the lack of international agreement on what is hazardous waste, making transboundary shipments difficult to monitor. The Basel Convention (discussed in more detail later in this chapter) comes closest to a universal definition by providing a list of regulated categories and characteristics of hazardous waste (Table 6.3).

Redressing the toxic past

Historically, most hazardous waste was simply dumped either into the closest stream or put in barrels and sent to the nearest dump. The discovery of 17,000 drums littered on a 7-acre site in Kentucky oozing hazardous waste onto the ground, led to national concern about similar 'Valley of the Drums' sites throughout the nation. Love Canal, a residential subdivision built on top of a chemical-waste burial site from the 1940s and 1950s tipped the balance on the public concern scale. Throughout the 1970s horror stories of abandoned waste sites and midnight dumpers who would take your hazardous waste for a fee and then dispose of it (only they knew where) continually made the evening news (Brown 1980b). Public outrage from the highly publicized events finally prompted Congress to action.

TSCA, RCRA, and Superfund

The first piece of important US legislation was the Toxic Substances Control Act signed into law in 1976. This legislation regulates chemicals to reduce public health and environmental exposures of chemicals and mixtures. The law requires industry to provide information (toxicity and

Table 6.3 Substances regulated under the Basel Convention.

Waste streams • clinical medical (hospital, dentist/doctors surgery, laboratories) • pharmaceutical, drugs, medicines with cyanides • waste mineral oils, waste with PCBs
Wastes with constituent parts of heavy metals (beryllium, copper, arsenic, zinc, cadmium, lead), asbestos, phenols, halogenated organic solvents, dioxin, organohalogens
Explosives
Flammable liquids
Flammable solids
Substances liable to spontaneous combustion
Substances that emit flammable gases when they come in contact with water
Oxidizing substances
Organic peroxides
Acute poisons
Infectious substances
Corrosives
Substances that emit a toxic gas when they come in contact with water or air
Toxins/carcinogens (including delayed and chronic effects)
Ecotoxins (leachate, substances that bioaccumulate)

Source: Mofson 1992.

environmental behaviour) about new chemical products that they plan to produce in a pre-manufacturing requirement. EPA reviews these applications, and if they are found to be environmentally benign approves them for use. If they are not found to be environmentally safe, EPA restricts their use. TSCA controls the manufacturing and distribution of new toxic chemicals, but has no authority over the thousands of chemical mixtures manufactured prior to the law's implementation in 1976.

In 1976 Congress passed another important law that controls wastes, or the by-products of industrial activities. The Resource Conservation and Recovery Act (RCRA) mandates strict controls over the management of hazardous wastes throughout their entire life cycle. It controls existing and future generators of hazardous waste through a cradle-to-grave tracking system for hazardous waste (where the waste originates to where it is ultimately disposed). RCRA not only identifies hazardous waste but has detailed reporting requirements and safety standards for generators and transporters as well. Design standards and operating procedures for facilities that treat, store, or otherwise dispose of hazardous waste are also part of RCRA. Only permitted facilities are allowed to handle hazardous wastes and the EPA has enforcement powers that can compel the owner of a hazardous-waste site to clean it up if it poses a significant health threat. RCRA covers more than 200,000 generators, 320 high-temperature incinerators, and in 1989 tracked more than 233 million tons of hazardous waste that was treated prior to disposal.

To clean up the country's toxic legacies, Congress passed the Comprehensive Environmental Response, Compensation and Liability Act (CERCLA), better known as Superfund in 1980. CERCLA provides the legal and financial mechanisms for coping with hazardous-waste site clean-up. Monies were set aside by Congress to facilitate the clean-up of hazardous-waste sites. Sites which pose the most serious public health and environmental threats are placed on the National Priorities List (NPL). A recent estimate places the potential number of sites from 130,000 to 425,000, and includes not only hazardous-waste facilities but underground storage tanks, underground injection wells, and federal civilian and military facilities (USGAO 1987). To determine which sites belonged to the NPL, the EPA created a national inventory of sites where hazardous wastes were stored, treated, disposed or released. The Comprehensive Environmental Response, Compensation, and Liability Information System (CERCLIS) contained 8000 sites in 1980, 22,000 in 1985, 27,000 in 1987 and 35,000 in 1991 (CEQ 1992). Only a small fraction of these will make the NPL list. In 1991 1207 sites were listed on the NPL including ten in Puerto Rico. New Jersey leads the nation in NPL sites with 108, followed by Pennsylvania (95) and California (88). Every state has at least one NPL site including Nevada (one), North Dakota and Mississippi (two apiece), and Wyoming and South Dakota (three each). Regionally, NPL sites are also concentrated in the Northeast, Great Lakes, and California (Fig. 6.2).

Fig. 6.2 National priority list (Superfund) sites. There are 1185 current sites with another 22 proposed during 1991 giving a total of 1207. Source: CEQ 1992.

Most of the contamination is from organic chemicals, solvents, and metals, seriously impacting groundwater, drinking water, and soils. The NPL sites are typically remnant manufacturing sites as well as old municipal landfills. Thus far, the Superfund programme has been much better at identifying and listing sites than actually cleaning them up. For example, by 1980 only 18 sites were certified 'clean' despite US $2 billion in federal funds. The costs of clean-up of the remaining sites is now estimated at US $1 trillion, not counting legal fees (Abelson 1992). This lack of effectiveness has lead to many critiques of the programme and frustrations by regulators, law-makers, and industry (Schweitzer 1991; Mazmanian and Morell 1992).

State programmes

CERCLA was designed to clean up hazardous-waste sites in conjunction with state-run programmes. In addition to NPL sites, individual states have identified more than 30,000 sites in need of remediation (Hall and Kerr 1991). California, Pennsylvania and Texas have been the most aggressive, listing more than 2300 sites apiece. Regionally, state and local determinations of hazardous-waste sites shows a slightly different spatial pattern than the NPL, with the Midwest showing the greatest concentration (Fig. 6.3). New Jersey has the largest number of Superfund sites, yet lists only 1300 additional sites in need of clean-up. Despite its prominence on the Superfund list, New Jersey actually trails in the toxic legacy race behind California, Pennsylvania, and Texas who have almost twice as many sites (Superfund and non-Superfund). North and South Dakota have the fewest locally identified hazardous-waste sites (58 and 72, respectively).

In 1983 New Jersey passed the Environmental Cleanup Responsibility Act (ECRA) that requires the owner or operator of any industrial facility and some commercial facilities such as gasoline stations to clean up any environmental contamination that may be on-site prior to the transfer or sale of that property. It does not matter if the use of the property stays the same or changes. In effect, industrial property cannot be bought, sold or leased within the state without certification that it is contaminant-free. In the autumn of 1991, 615 sites were identified as containing multiple sources of contamination, or effecting multiple media (air, soil or groundwater). These complex sites including 108 on the Superfund list, do not represent the totality of remediation efforts in the state. However, they do provide some measure of the effectiveness of New Jersey's hazardous-waste clean-up effort, the most aggressive programme in the country. The majority of sites are concentrated in the northern industrialized core of the state in Middlesex (73), Bergen (53), and Hudson (55) counties. It is interesting to note that the majority of Superfund sites are also located in these counties (Fig. 6.4), which coincidentally have the highest population densities as well.

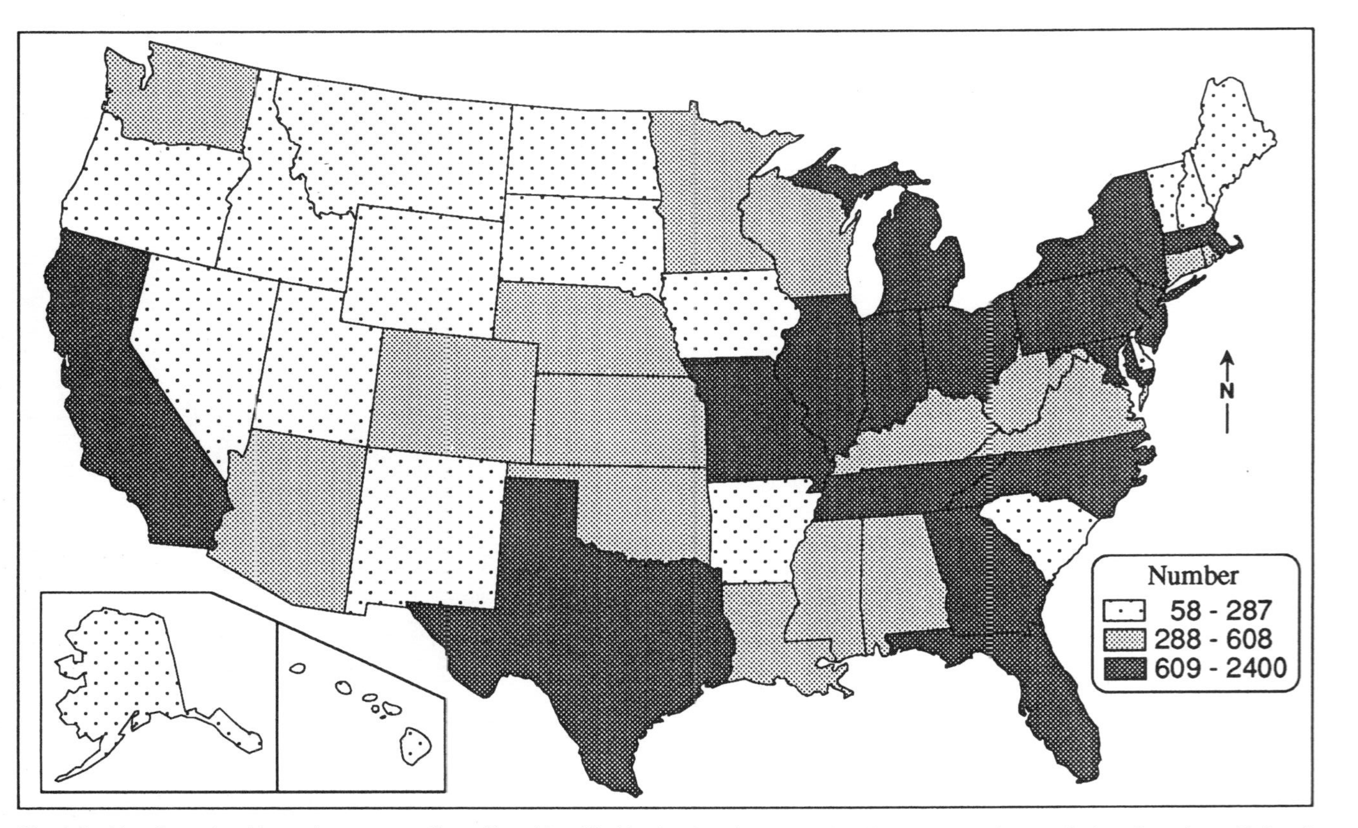

Fig. 6.3 Non-Superfund hazardous-waste sites. Sites identified by local and state authorities in need of remediation. Data from Hall and Kerr 1991.

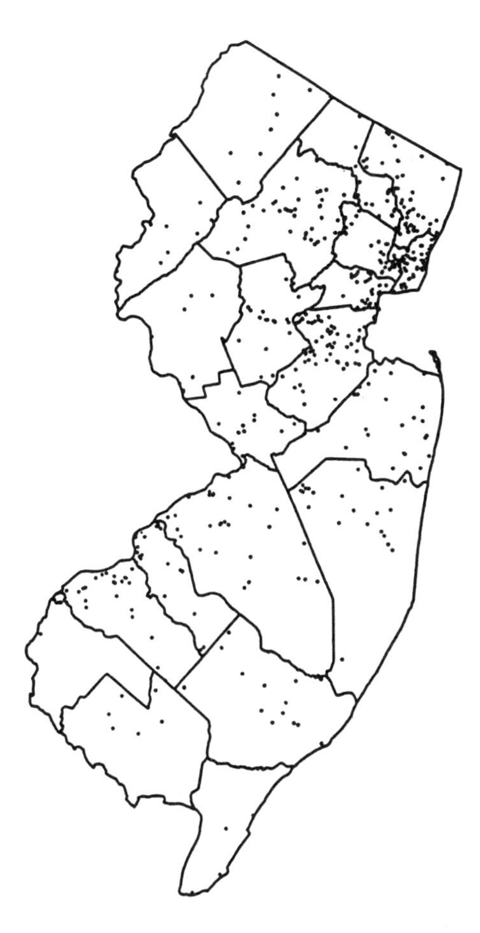

Fig. 6.4 New Jersey's hazardous waste sites, 1991. More than 615 sites have been identified in the state including the 108 on the national Superfund list. Source: NJDEPE 1991.

Risk burdens

An important element in examining the hazardous-waste issue is the distribution of risks and hazards. Where and who are the risk generators? Furthermore, who bears the burdens of risk and why?

Risk generators

There are two primary sources of hazardous waste: commercial/industrial processes and the military. Given the paucity of data, most estimates of hazardous waste only include commercial and industrial generators. The role of the military in hazardous-waste production has been ignored by most nations or simply listed as classified information.

Commercial/industrial

We have already mentioned the US distinction as the world's largest producer of hazardous waste. Yet, there are significant differences within the US in where hazardous wastes are generated and volumes of waste discarded. In 1985 (there have been no comprehensive surveys since then) there were more than 21,000 facilities that generated hazardous waste (Hall and Kerr 1991). California had the most (3972) followed by Pennsylvania (2607) and Texas (2450) (Fig. 6.5). Yet, simply looking at the number of facilities does not give any indication of the volume of waste generated at each site. California, for example, may have quite a number of smaller facilities generating relatively low levels of waste, compared to Louisiana which has fewer but larger facilities. Another measure is the likely population at risk or the number of facilities per capita. Using this measure, the greatest risk potential is found in Rhode Island (40.3/100,000 people), Vermont (22.1/100,000 people), Pennsylvania (21.9/100,000 people), New Jersey (19.1/100,000 people), Oregon (17.8/100,000 people). The lowest risk potential is found in South Dakota (0.6/100,000), North Dakota (1.3/100,000), and Alaska (1.6/100,000).

Approximately 250 million tons of hazardous waste were generated in 1985, the majority of it in southern states. On a per capita basis, Tennessee is the largest producer (Fig. 6.6), averaging just under 7 tons per person per year from 556 generators like Eastman Kodak, General Electric, Whirlpool, Union Camp and Chemical Products Corporation (Hall and Kerr 1991). Georgia is next with 6.2 tons per capita from 330 generators, and West Virginia is third also with 6.2 tons per person from only 57 generators.

In addition to manufacturing and industrial producers, two other sources of risk from hazardous waste are worth mentioning. The first is hazardous-waste management facilities. These commercial facilities, now numbering around 5000 are regulated under RCRA and treat, store, dispose or somehow 'manage' hazardous waste. Texas leads the nation in the number of facilities (1153) followed by Pennsylvania (464), California (348), Illinois (295), and New Jersey (284). Not surprising the facilities are spatially concentrated in industrialized Northeast and Midwest states (Fig. 6.7). The fewest number of hazardous-waste management facilities are in the intermontane west. On a per capita basis, the greatest risk is found in Texas, Connecticut, and Pennsylvania, while the least is found in North and South Dakota and Nebraska.

Finally, the last source of hazardous-waste risk is transportation accidents since much of the hazardous-waste material is continually on the move from producer to consumer, consumer to disposer, and from producer to commercial waste-management facility. These transfers occur within and between states. During the 1980s, more than 5800 incidents involving the transport of hazardous materials by truck, rail, ship or

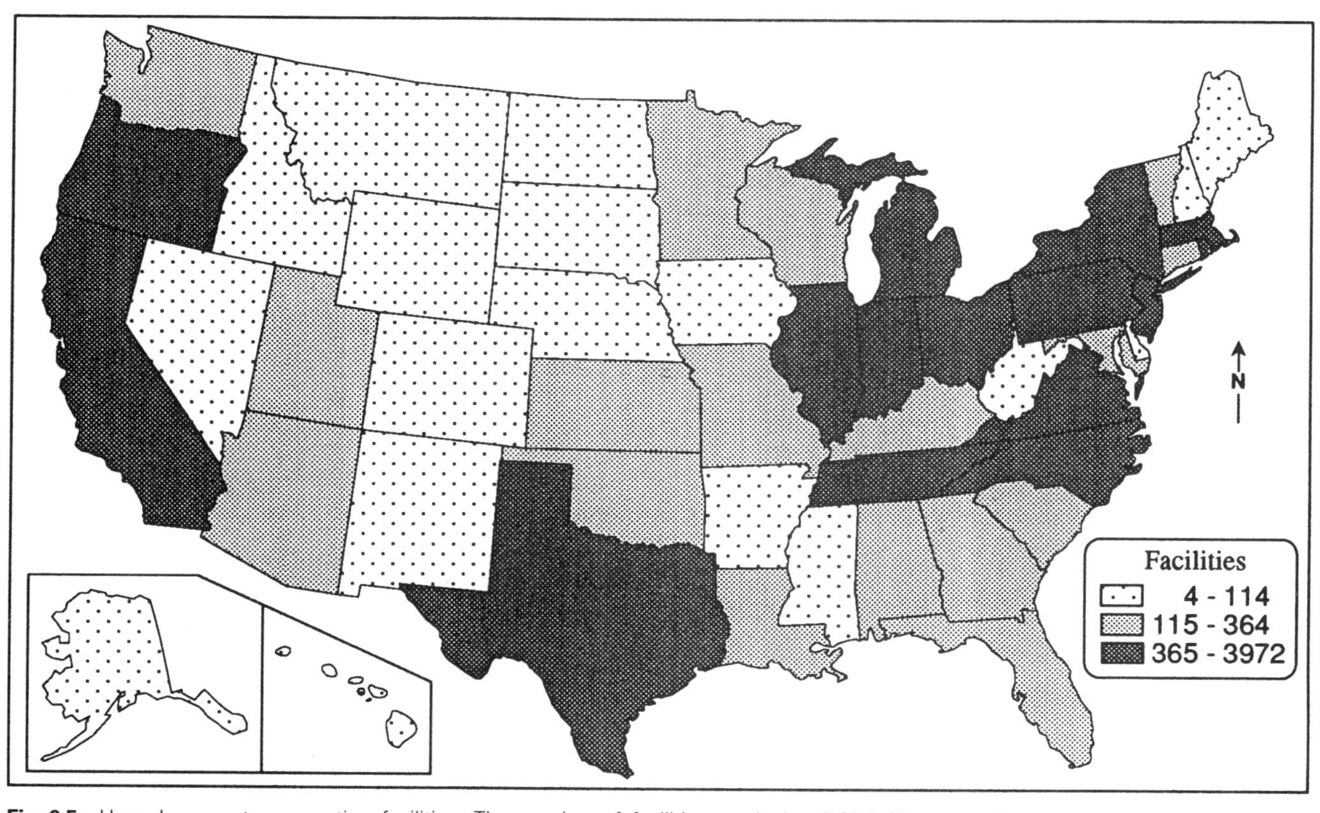

Fig. 6.5 Hazardous-waste generating facilities. The number of facilities producing RCRA (Resource Conservation and Recovery Act) regulated hazardous waste in 1985. The median number of facilities per state is 191. Data from Hall and Kerr 1991.

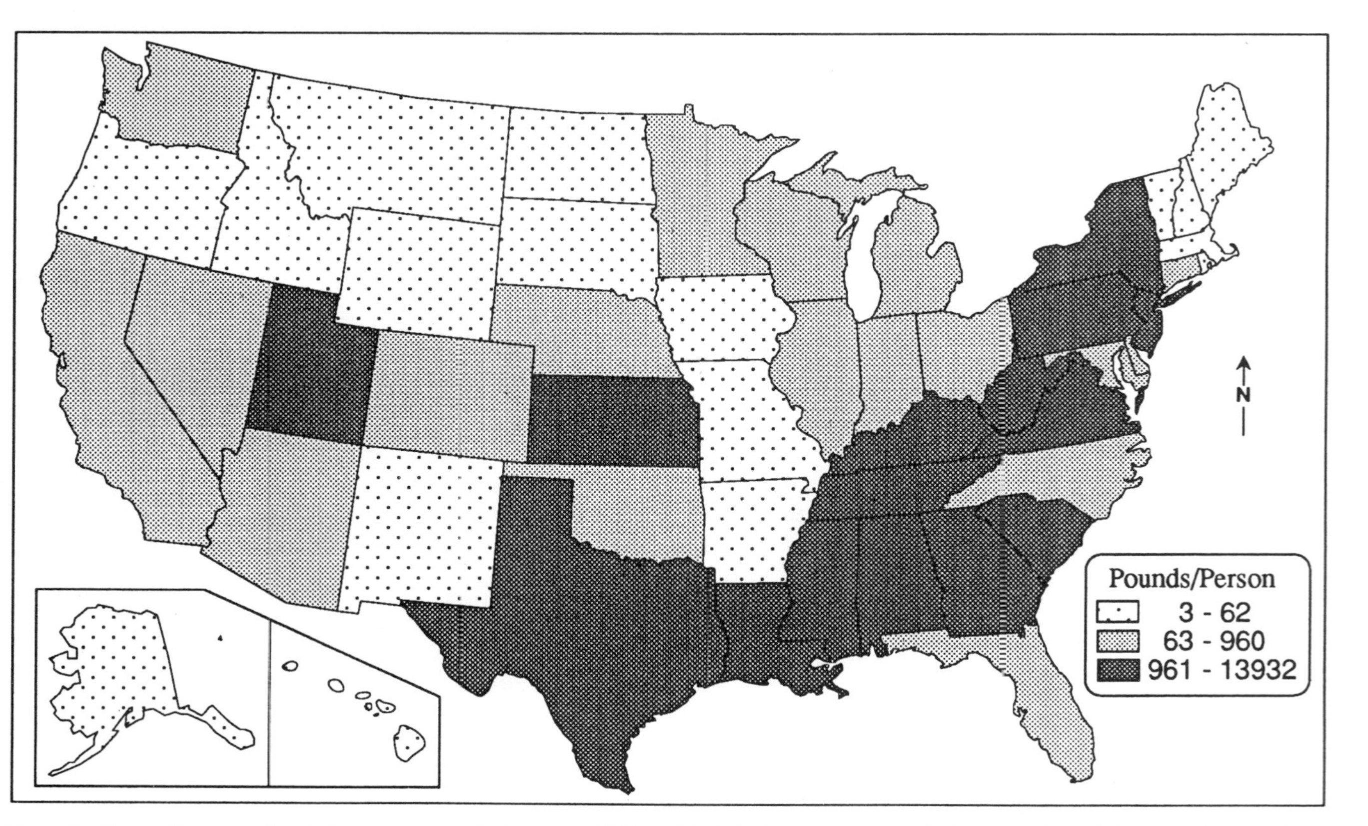

Fig. 6.6 Per capita generation in lbs per person. Data are for 1985 and include hazardous waste-laden water in addition to solid hazardous wastes. Data from Hall and Kerr 1991.

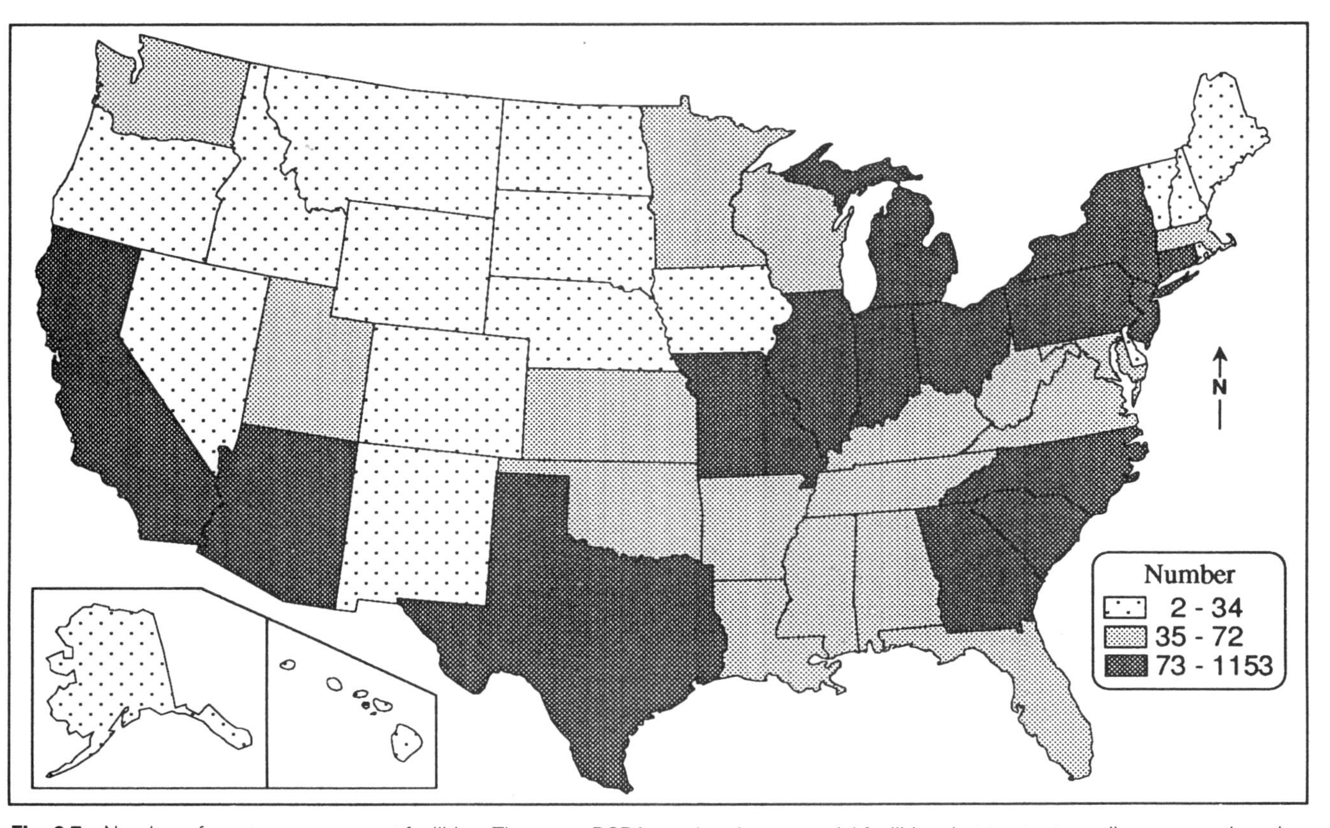

Fig. 6.7 Number of waste-management facilities. These are RCRA regulated commercial facilities that treat, store, dispose, or otherwise manage hazardous wastes. Data from Hall and Kerr 1991.

airplane occurred yearly, most of them highway incidents. By 1989, the number of incidents began to rise reaching more than 9000 by 1991 (Table 6.4). During 1991 the largest number of incidents took place in California, Pennsylvania, Ohio and Illinois, the fewest in Hawaii, North Dakota, and Rhode Island. In addition to concentrations in the industrialized East and Midwest, transporation incidents are also clustered in the Southeast, Texas-Louisiana, and California (Fig. 6.8). Nearly 40% of the incidents involved corrosive materials while another 37% involved flammable/combustible liquids. Damage estimates rose between 1986 and 1991 topping more than US $37 million by 1991. Injury and death statistics were more variable.

Table 6.4 Trends in hazardous materials incidents.

Type of incident

Year	Water	Highway	Railroad	Air	Total
1987	16	4952	888	163	6019
1988	16	4906	1020	172	6114
1989	11	6036	1195	188	7430
1990	7	7256	1273	294	8830
1991	13	7584	1130	294	9021

Injuries/damages

Year	No. major injuries	No. minor injuries	No. deaths	Damages ($ million)
1987	21	310	10	23.4
1988	36	135	19	21.3
1989	35	295	8	26.1
1990	42	376	8	32.1
1991	26	410	10	37.6

Source: USDOT 1992.

Military

The US military is the largest generator of hazardous waste in the world producing more than 450,000 tons of hazardous waste and 850,000 tons of contaminated wastewater in 1989 (Seigel, Cohen and Goldman 1991). The military's toxic legacy includes 14,401 'hot spots' at 1579 facilities in the US alone, an average of nine sites per facility. By state, California has the most (1713) while Vermont has the fewest. Regionally, military generators are concentrated in the western and southern states (Fig. 6.9).

The US weapons complex run by the Department of Energy also produces an assortment of hazardous waste (Table 6.5) in addition

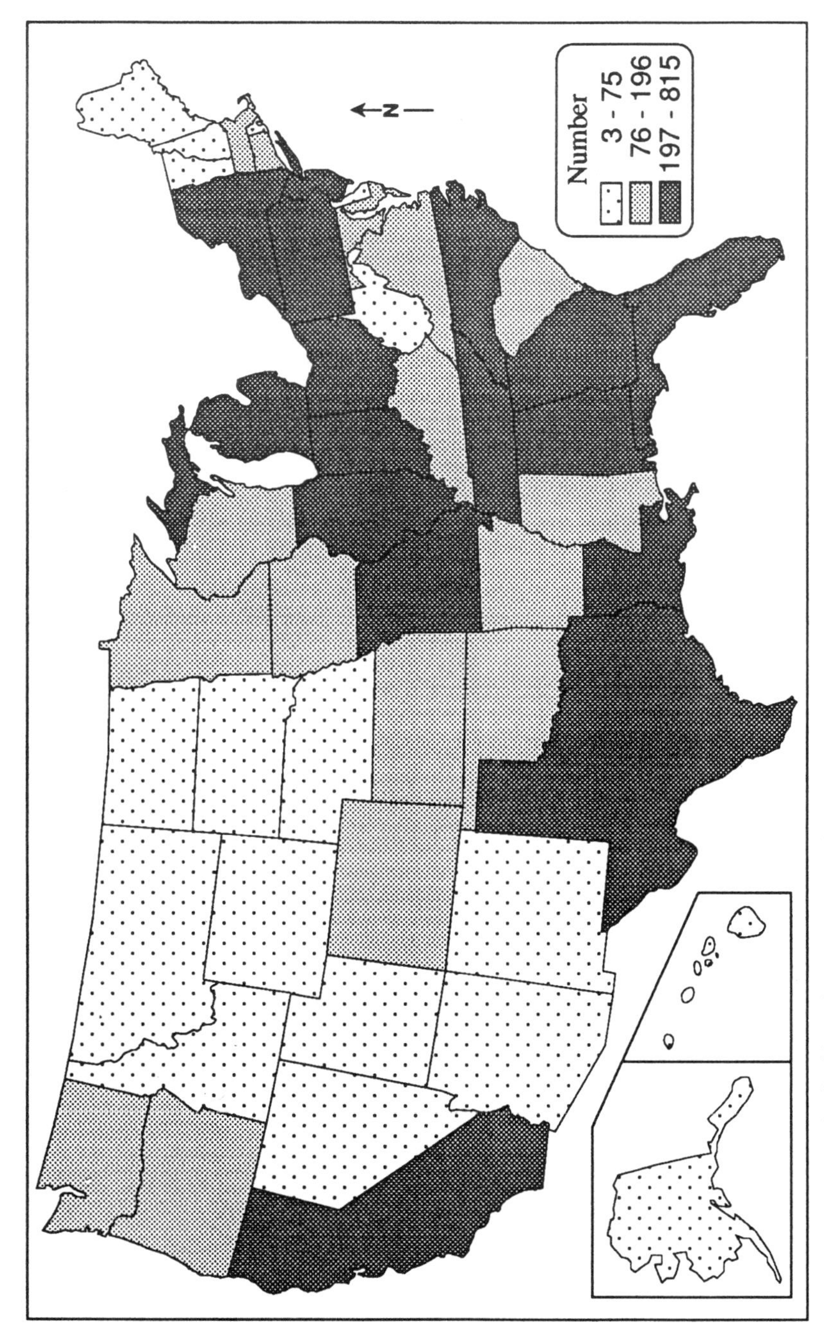

Fig. 6.8 Transportation incidents in 1991. The total number of incidents involving hazardous materials was 9021 for the year. The median number of incidents was 109.5. Data from USDOT 1992.

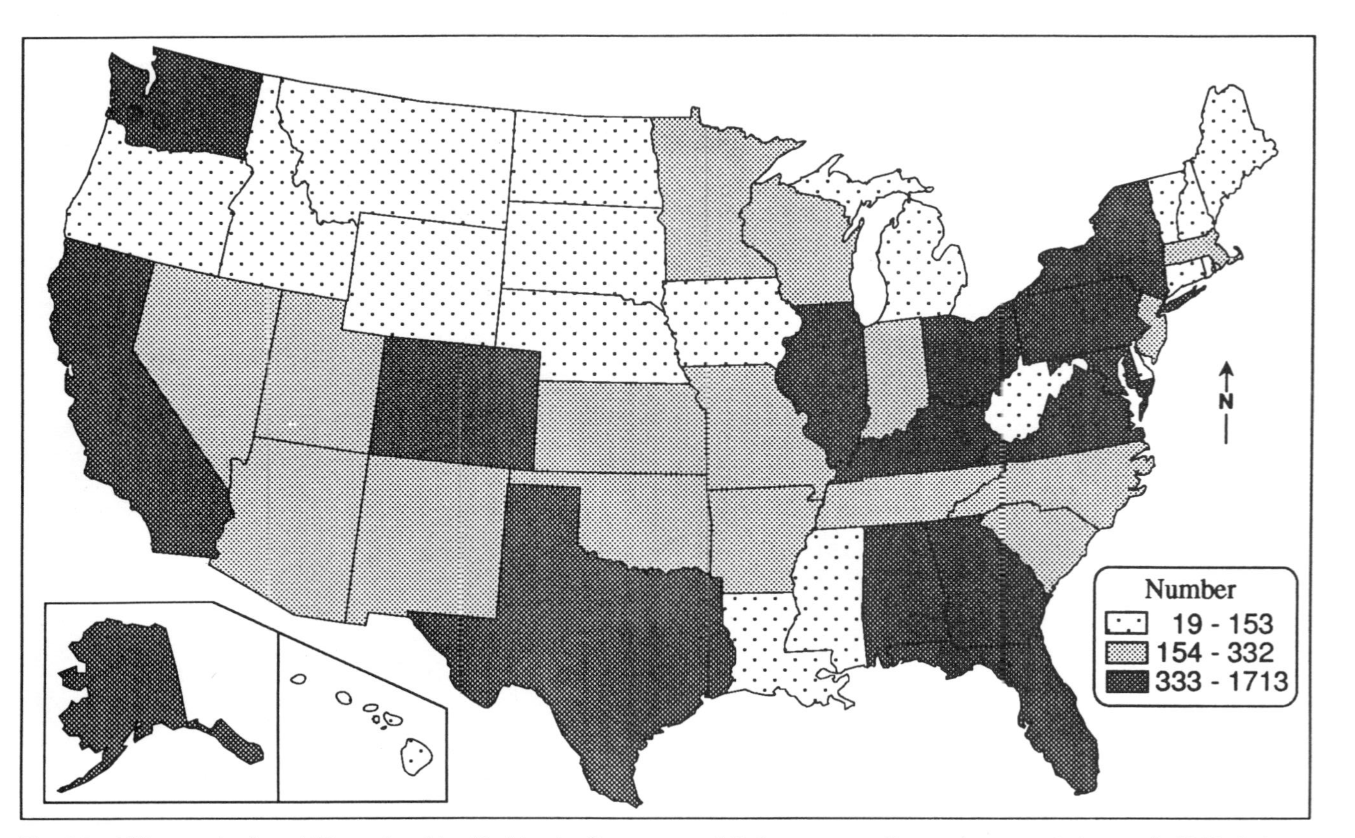

Fig. 6.9 Military toxic sites. Military sites identified by the Department of Defense as needing environmental clean-up in 1989. At least 100 of these are already on the EPA's Superfund list. Source: Data from Seigel, Cohen and Goldman 1991.

Table 6.5 Contaminants at DoE weapons facilities.

Inorganic contaminants
Radionuclides: Americium-241; Cesium-134, 137; Cobalt-60; Plutonium-238, 239; Radium-224, 226; Strontium-90; Technetium-99; Thorium-228, 232; Uranium-234, 238.
Metals: chromium, copper, lead, mercury, nickel.
Other: cyanide.

Organic contaminants
Benzene, chlorinated hydrocarbons, methylethyl ketone, cyclohexanone, acetone, polychlorinated biphenyls (PCBs), selected polycyclic aromatic hydrocarbons, tetraphenylboron, toluene, tributylphosphate.

Organic facilitators
Aliphatic acids, aromatic acids, chelating agents, solvents, diluents, and chelate radiolysis fragments.

Mixtures of contaminants
Radionuclides and metal ions; radionuclides, metals, and organic acids; radionuclides, metals, and natural organic substances; radionuclides and synthetic chelating agents; radionuclides and solvents; radionuclides, metal ions, and organophosphates; radionuclides, metal ions, and petroleum hydrocarbons; radionuclides, chlorinated solvents, and petroleum hydrocarbons; petroleum hydrocarbons and polychlorinated biphenyls; complex solvent mixtures; complex solvent and petroleum hydrocarbon mixtures.

Source: US Congress 1991, p. 24.

to radioactive wastes. The primary weapons production occurs at 15 sites (Fig. 6.10) and involves research and design, materials production, actual manufacturing of weapons, warhead testing, and waste disposal (Table 6.6). The most contaminated sites include the Hanford Plant, Savannah River, Fernald, Rocky Flats, Oak Ridge, and the Mound Plant, all with on- and off-site contamination of soils and sediment, groundwater, and surface water. Organic solvents like trichloroethylene (TCE) and perchloroethylene (PCE) are present at most DoE sites and present the largest potential cost in remediation efforts. Remediation cost estimates range from US $100 to US $360 billion for these 15 sites alone! (Abelson 1992).

Toxic politics

Who bears the burdens of the risk is directly related to the politics of place. There is of course an uneven development of industrial activities setting the stage for local and regional disparities in the location of industrial activity. There is also a social stratification of residents (largely based on wealth). At the local level then, you generally find lower income (and often minority) communities abutting industrial facilities. At the national level, Bullard (1990) has demonstrated that the South bears a

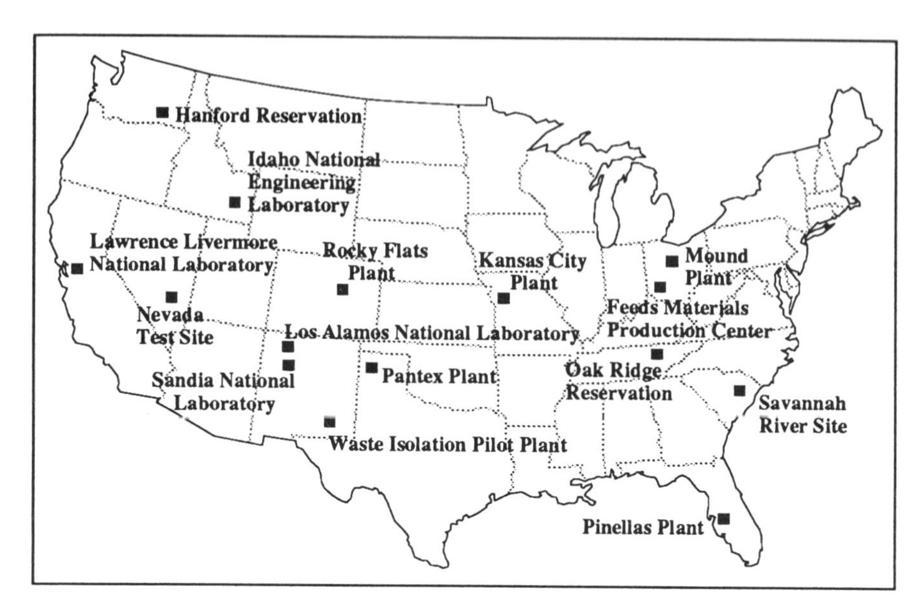

Fig. 6.10 US Department of Energy's weapons complex. Source US Congress 1991.

disproportionate burden of hazardous-waste landfills. In both cases, poor and minority residents lack the economic power and political clout to prevent the influx of unwanted and risky technologies. They also have few if any opportunities to relocate to less hazardous locations, thus becoming victims.

The only opportunity many of these victims have is to seek redress by making the problem public, organizing the community, and mobilizing political constituencies (Reich 1991; Shulman 1992); to fight back against big business and big government. As individuals and communities are directly affected by the risk and hazards, they begin to mobilize. We have seen instances of this in minority communities with the rise of the environmental justice movement (see Chapter 4).

One of the most well-known examples is the role of citizens in bringing Love Canal to public attention (Levine 1982). Spearheaded by a concerned housewife, Lois Gibbs, residents in the Niagara Falls community noticed a number of birth defects, miscarriages, and cancer deaths in their neighbourhood during the mid-1970s (Gibbs 1982). Unbeknown to them, Hooker Chemical Company had used the area 30 years earlier as a dump site, simply covering over abandoned waste drums in the drained canals. As Niagara Falls grew and new housing built, the dump site was forgotten and eventually redeveloped into a nice residential area. Then barrels started leaking into basements and popping up from the soil after heavy rains. Children became sick and parents worried. The residents of Love Canal mobilized to bring these accounts to the attention of public health officials, who were at first extremely sceptical of their accounts of chemical exposure. Eventually the true extent of the contamination was realized and between 1978 and 1979 more than 100 homes were evacuated (the government purchased the homes and relocated the residents) and the area was completely cordoned off. By 1990, in an interesting twist, the US government began selling the Love Canal homes, declaring the neighbourhood clean and habitable.

Love Canal became a symbol of America's toxic legacy, and Lois Gibbs the champion of citizen outrage, empowerment, and unwillingness to

Table 6.6 Weapons facilities and medium of environmental contamination.

Type of facility	Location	Contamination on-site	Contamination off-site
Weapons research/ design	Los Alamos Natl. Lab.	S, GW	
	Sandia Natl. Lab.	S, GW	
	Lawrence-Livermore Natl. Lab.	S, GW, Se	GW
Materials production	Hanford Plant	S, GW, SW, Se	SW, Se
	Savannah River Site	S, GW, SW, Se	SW, Se
	Feed Materials Production Center (Fernald)	S, GW, SW, Se	GW, Se
	Idaho Natl. Engineering Lab.	S, GW, Se	S
Weapons manufacturing	Rocky Flats Plant	S, GW, SW, Se	SW, Se
	Oak Ridge Reservation	S, GW, SW, Se	SW, Se
	Mound Plant	S, GW, SW, Se	GW, SW, Se
	Pinellas Plant	GW	
	Kansas City Plant	S, GW, Se	
	Pantex Plant	GW	
Warhead testing	Nevada Test Site	S, GW	
Waste disposal	Waste Isolation Pilot Project		

S = soil, GW = groundwater, SW = surface water, Se = sediment.
Source: US Congress 1991, p. 26.

bear the risks. The tactics and lessons of Gibbs' grass-roots organizing are replicated time and again throughout the country. She continues her efforts to empower local citizens against industrial hazards as the founder of the Citizens' Clearinghouse for Hazardous Waste.

Where does hazardous waste go?

Until recently the common practice for hazardous-waste management was discarding it either on-site or having it hauled off-site by a waste hauler, often one with a dubious record. Things have changed and now that hazardous waste is more tightly regulated, midnight haulers and shady operators are less common, but they do still exist. In fact, hazardous waste has become such a problem for generators, that an entire new industry has developed in response to it.

Hazardous-waste management firms provide a litany of services ranging from assessments on the types of waste requiring action, to treatment, recycling, storage, and disposal. In fact, business is booming. The commercial off-site treatment and disposal market grew to US $3.5 billion in 1991, an increase of US $0.5 billion from the previous year (Chemical Week 1991a). They also provide technical assistance on methods to minimize the production of wastes in the first place. As we shall see in the next section, hazardous waste is often viewed as a marketable commodity that is imported and exported. One country's or state's waste becomes another's source of revenue.

Management alternatives

Broadly speaking, there are four alternatives in coping with the mounting problem of hazardous wastes. These include waste minimization, resource recovery/recycling, treatment, and disposal.

Waste minimization

The most effective type of hazardous-waste management is not to produce it in the first place. Waste minimization is now in vogue as many industries are modifying their industrial processes to reduce hazardous-waste generation as the costs of disposal and treatment rocket. While not universally accepted, waste minimization at the source (industry, consumers) is perhaps the most preferred method of management by regulators and the public. Even some in industry are beginning to champion this approach as well. The Toxic Release Inventory discussed in Chapter 5 provides one indicator of pollution-prevention trends where toxic manufacturing emissions are steadily declining through a combination of voluntary and mandatory reductions. The 1990 Pollution Prevention Act establishes the principle as national law. Management priority must be given to pollution (including hazardous waste) prevention and reduction at the source. The law also establishes a hierarchy for management alternatives: prevention, recycling, treatment, and only as a last resort, disposal.

Resource recovery/recycling
Another option in managing hazardous wastes is to recycle the wastes within the industry or transfer them to another industry who can use the materials. Reprocessing the waste for energy recovery or materials recovery such as extracting valuable basic ingredients in the waste like heavy metals and catalysts is now being tried by a number of industries.

Treatment
A third alternative is the treatment of hazardous waste to render it less hazardous. This includes a number of different processes ranging from physical treatment such as distillation and sedimentation, to chemical treatment (fixation to solids, oxidation), to biological treatment (composting, activated sludge). Treatment invariably reduces the volume of hazardous waste in need of disposal.

Disposal
Traditionally, disposal has been the preferred option for managing hazardous waste. Disposal occurs in three forms: landfills, incineration, and exporting waste from the immediate area. The placement of hazardous waste into landfills (both regulated and non-regulated) has a long history that we've already discussed. A variant on this land-based method is the injection of hazardous wastes into deep wells (below the existing water tables) as a management option. Incineration of wastes is also technologically feasible, but local community resistance limits its usefulness, particularly in the US. There are at-sea incineration programmes, notably the Dutch vessel *Vulcanus* that incinerates organic wastes offshore, but again these are extremely limited.

Current practices

Historically, the vast majority of hazardous waste in the US was landfilled. In 1988, 64% or 4.2 million tons were managed in this manner, a threefold increase from 1983 (Table 6.7). Yet by 1990, landfilling of hazardous waste was a less viable management option (Fig. 6.11). Over time, the use of management methods changed. For example, in 1988 resource recovery including the recovery of energy and materials lagged behind landfilling as the preferred method. By 1990, however, resource recovery (energy and materials) became the most dominant management alternative with 45% of the hazardous waste being handled in this way. Treatment options to reduce the hazardousness of the material have also declined since the early 1980s. Deep-well injections declined from 1983 to 1988 as regulatory

Table 6.7 Trends in hazardous-waste management options (millions of tons).

Method	Year 1983	1987	1988	1990
Incineration	0.18	0.48	0.67	1.20
Treatment (biological/ chemical/physical)	0.90	1.00	0.15	0.19
Solidification	0.06	0.07	–	0.05
Resource recovery	0.30	0.32	1.20	2.16
Landfill	1.00	3.00	4.20	0.17
Deep-well injection	0.75	0.28	0.39	0.74

Sources: Chemical Week 1988; Rotman 1990, 1992.

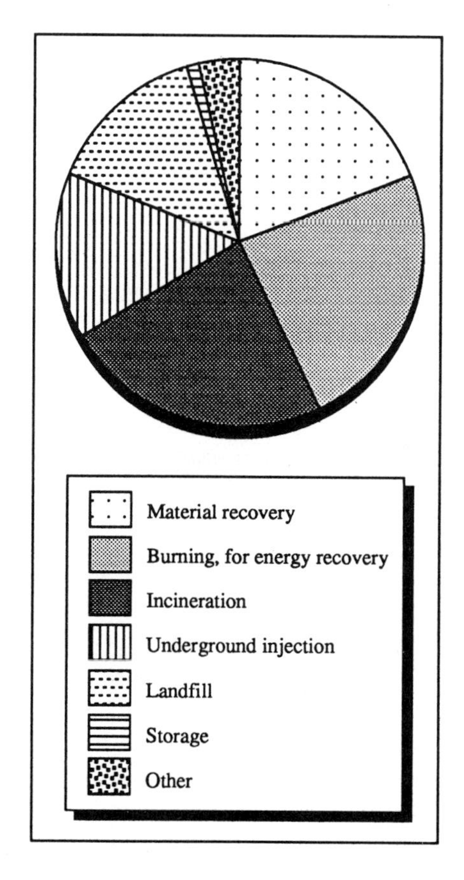

Fig. 6.11 Where hazardous waste goes. Most of the hazardous waste in the US in 1990 was incinerated or burned for energy recovery. Source: Rotman 1992.

barriers (the Hazardous and Solid Waste Amendments, 1984) were put in place that actively discouraged landfilling and deep-well injection as disposal options. By 1990, however, deep-well injections were again on the rise. As the hazardous-waste rules under the Resource Conservation and Recovery Act become more restrictive, the costs of hazardous-waste management increase and the options become more limited.

At the same time, there has been an increase in hazardous-waste management firms, sparking a new industry to clean up the old. Between 1971 and 1980 revenues grew from US $60 million to over US $315 million, and by 1984 the industry generated US $1 billion in revenues. In 1991 estimates reached more than US $7.2 billion in net sales from the 13 largest firms (Chemical Week 1992). The largest growth sector in the off-site commercial hazardous-waste market is in incineration which has become one of the most used options in the early 1990s. In 1988 there were 309 commercial incinerators, processing 69,143 tons of hazardous waste (Table 6.8). By late 1991 16 commercial incinerators were proposed adding an additional capacity of 973,000 tons per year (Chemical Week 1991b).

Table 6.8 Hazardous-waste treatment facilities, 1988.

Method	Company-owned		Commercial	
	Tons	No. facilities	Tons	No. facilities
Incineration	611 806	105	69 143	309
Chemical treatment	40 212	8	4528	17
Physical treatment	83 983	4	130	2
Biological treatment	11	3	5187	3
Thermal	3614	3	1231	3
Solidification	246	2	3804	12
Combination	12 419	7	5230	10
Not specified	994	4	936	8
Total	753 285	136	90 189	364

Source: Rotman 1990, p. 40.

As the costs of hazardous-waste management increase, industry is attempting (with federal prompting in the form of the Pollution Prevention Act) to minimize the volume of hazardous waste produced. To reduce the costs of disposal and future problems, industries that generate hazardous waste are now changing their manufacturing processes to minimize hazardous waste or reduce it completely. For example, Chevron USA, Inc. established their Save Money and Rescue Toxics (SMART) programme. This company-wide initiative encourages individual sectors within the firm to reduce toxics pollution by establishing a goal of 65% reductions. Programmes and managers are evaluated on the basis of their meeting these goals, with significant rewards for those who achieve them. In the first year, Chevron USA reduced their hazardous-waste generation from 135,000 tons to 76,000 tons, thereby saving US $4 million in waste-disposal costs as well! (CEQ 1991).

Despite these initiatives, hazardous-waste generators still must somehow cope with increasing volumes of hazardous waste and mounting costs of disposal. Instead of handling the wastes in-house, generators often contract with waste-management services to haul the waste off-site. Where it goes after that, out-of-state, out-of-country, is rarely considered by the generator. It simply is not their problem anymore.

Transboundary trade

The export of hazardous waste between regions and countries regularly makes the evening news with stories on midnight dumpers, errant barges, and the odysseys of poison ships. Perhaps no other event has highlighted the scandals of hazardous-waste mismanagement in the international arena than the ill-fated voyage of the *Karin B.* In 1988 the *Karin B.*, a modern-day equivalent of *The Flying Dutchman*, began her saga from Livorno, Italy. The West German ship was carrying 210,000 tons of toxic waste; its destination Koko, Nigeria, where the cargo was to be landfilled near a remote Nigerian village. After protests by environmental groups, the cargo was refused and the *Karin B.* was sent back to sea. Calling at the Canary Islands, Spain, England, and France, and refused entry at each port, the ship eventually returned to Livorno, its cargo intact. The cargo was unloaded (at a cost of US $11 million) and returned to the generators (two Italian chemical firms) for ultimate disposal (Center for Investigative Reporting and Moyers 1990). Ironically, the wastes eventually ended up in the United Kingdom where they were incinerated (European Chemical News 1988; Greenhouse 1988).

The *Khian Sea* is another example. The freighter carried 15,000 tons of toxic ash from Philadelphia's municipal incinerator, bound for Haiti where the ash was dumped on the beach as fill until the Haitian government halted it in early 1988. After two years, two name changes, and a voyage touching four continents, the *Khian Sea* (now named the *San Antonio*) appeared in Singapore without its cargo, presumably dumped somewhere in the Bay of Bengal. As Moyers writes, 'The Homeric odyssey that had lasted more than twenty-seven months, touched four continents, and set off a series of international furors had in the end involved no more than one month's worth of incinerated waste from US city' (1990, p. 30).

Many other examples of toxic shipments are less well known (Table 6.9) yet illustrate the nature of the international 'trade' in hazardous waste. There are many more examples where the waste schemes were ultimately rejected at the last moment or whose fate is unknown (Greenpeace 1990).

Table 6.9 Selected recent incidents of known toxic dumping.

1) 1984–1986, Canna & Dan, Benin
Discovered in 1989 (via groundwater contamination) that radioactive waste from former USSR was illegally dumped below an airfield.

2) 1988, Abomey, Benin
Agreement with France (US $20 million plus a 30-year promise of economic assistance) in exchange for nuclear-waste burial. Agreement with Sesco Ltd to accept 5 million tons of toxic and mixed wastes. Benin paid US $2.50 per ton of waste (US $12.5 million total) while Sesco clients (western Europe and the US) pay upwards of US $1000 ton.

3) 1988–89, Congo
Congolese government agreed to contract with Export Waste Management (New Jersey) to accept 20,000–50,000 tons of waste (including pesticide residues and industrial sludge) per month from the US and West Germany. After a few shipments were received, the government cancelled the deal and arrested three Congolese officials involved in setting it up.

4) 1988, Djibouti (city), Djibouti
Initial destination of 2400 tons of toxic-waste drums from Italy (collected by Jelly Wax). Shipment was refused, transported to Venezuela and Syria (where it was also refused), and then back to Italy. The cargo was finally sent to the UK for incineration.

5) 1987–1989, Lambarene, Gabon
Nuclear-waste dumping agreement with Denison Mining Corporation of Canada. Measurable levels of radioactivity found in water samples in Ogooue River which borders the city. Accepted radioactive waste from uranium mining companies in Colorado.

6) 1988, Koko, Nigeria
Destination of 4000 tons of toxic chemicals (PCBs, dioxins, asbestos) and radioactive wastes from Italy carried by the *Karin B*.

7) 1991, Lusaka, Zambia
A donation of beef from Czechoslovakia was found contaminated with radioactive toxins. Beef was buried in a deep hole, 40 miles east of Lusaka and covered with concrete slab. Despite the risk, villagers initially tried to chip away concrete to get at the free food.

8) 1978–1987, Zimbabwe
An unknown quantity of toxic waste from US armed forces and defense agencies (Colbert Brothers, waste haulers) was passed off as pure dry-cleaning solvent and sold to various companies in the country.

9) 1988, South Africa
Nearly 120 drums of mercury-contaminated waste has been shipped to Natal from the US (American Cyanamid of New Jersey) since 1986. A total of 32,400 lbs have been received.

10) 1989, Goanives, Haiti
Initial destination of the *Khian Sea* carrying several million tons of Philadelphia's toxic incinerator ash containing high levels of cadmium, mercury, arsenic, and dioxins. Originally 'sold' as fertilizer.

11) 1986, Tecate, Mexico
An unlicensed recycling company based in Tijuana paid by several US companies to handle their toxic wastes. Nearly 10,000 gallons of hydrocarbons and other toxic wastes were dumped, threatening groundwater supplies.

12) 1987–1991, Colon, Panama
Close to 7 million tons of flyash from US power plants were used as pavement materials for roads. International Energy Resources (US) proposed building an incinerator to burn New York City's garbage (up to 9000 tons per day), but the project was rejected.

13) 1987, Venezuela
More than 2000 tons of highly toxic wastes from Europe were offloaded at Puerto Cabello.

14) 1990– , Johnson Atoll
The island is currently being used by the US military to incinerate chemical weapons. Close to 100,000 nerve-gas shells shipped from Okinawa and other Pacific bases are on-site for the one incinerator. Nine more incinerators are planned to handle the additional 300,000 shells from other US military bases (mostly in Europe).

15) 1988, Hong Kong
Plastic wastes from Germany (135 tons worth) were burned in the local waste incinerator.

16) 1986, Japan
Between 2500–3500 metric tons of flammable and hazardous waste were imported from Formosa Plastics, Port Comfort, Texas. The objective is to recycle the wastes into feedstocks.

17) 1988, Lebanon
The Italian firm, Jelly Wax, shipped 2400 tons of hazardous waste to ports outside of Lebanese governmental control.

18) 1983, Belgium
A West German firm illegally dumped 27,000 tons of toxic waste into an abandoned quarry, paying US $15 per ton for the disposal including transportation costs. Comparable disposal in West Germany (US $30 per ton) or France (US $20 per ton) meant a saving of US $500,000.

19) 1988, Romania
More than 4000 tons of hazardous waste containing PCBs from Italy, Netherlands, and West Germany were discovered in an illegal dump near Sulina. The wastes were leaching into the Danube delta.

20) 1988–1989, Poland
Two Austrian firms exported 10,000 barrels of hazardous waste to Poland that were labelled as recyclable varnish and solvent residues. The barrels contained cyanamide, highly chlorinated solvents, and PCB-contaminated oils. Nearly 1000 barrels have been returned to Austria.

21) 1979– , Germany
Since 1979, the Schoenberg dump in the former East Germany has accepted more than 1 million tonnes of hazardous waste annually from many European countries. In 1988 1 million tonnes of toxic and municipal waste was imported from the former West Germany, 50,000 tonnes of toxic waste each from Austria and Italy, 35,000 tonnes of toxic waste from the Netherlands, and 4000 tonnes of toxic waste from Switzerland.

Sources: Greenpeace 1990; Center for Investigative Reporting and Moyers 1990; and various national and international newspaper reports.

Waste brokers

The hazardous-waste business is populated with sleazy brokers, intermediaries, obliging shipping firms, and nefarious ghost companies. Some of the most well-known international brokers include InterContract (Switzerland); Jelly Wax (Italy); Sesco Ltd (Gibraltar); Blauwerk (Liechtenstein); Bulk Handling (Norway); West Export Management (US); and Scoot Corporation (US). Jelly Wax was involved in the *Karin B.* fiasco among others. Unfortunately, developing nations are ideal locations for dumping wastes—they have vast tracts of unused land, public opinion on the dangers is non-existent, and government authorities can either turn an unconcerned eye or be provided with sufficient monetary incentive to look the other way.

The waste brokers collect and store waste from reputable companies at a fee. They choose a poor country, most frequently one with a large foreign debt. The brokers promise hard currency in exchange for dumping privileges. The brokers periodically promise longer-term infrastructure developments such as backfilling or roadbed construction or package deals: you take the waste and I'll provide discounts on other needed products like fertilizers, machinery, pesticides. The transport of the hazardous waste is carried by ships registered in countries with few trade regulations. Finally, and most importantly, wastes are often mixed (diluting the hazardous-material content) to bypass regulations, or are intentionally mislabelled as products for reuse. While clearly illegal, calling waste products by another name is almost impossible to detect. The waste brokers exist because industries are willing to pay vast amounts of money to get rid of their wastes, and poor countries are willing to accept them in exchange for monetary compensation, often disregarding the risks in doing so. Foreign exchange, for many of these debt-ridden countries, is often more important than the health of the citizens or their environment. The international trade in hazardous waste is booming especially when you consider that it costs upwards of US $200–300 per ton to dispose of hazardous waste domestically, while some developing nations charge less than US $40 per ton, a substantial saving (Chepesiuk 1991).

Domestic highways

In the United States, the EPA is notified of all transboundary shipments of hazardous waste, be they between states or between the US and other countries. Under the provisions of the RCRA's cradle-to-grave tracking system, hazardous waste is controlled from the point of generation to its ultimate disposal, in other words, through its entire life cycle. Generators, transporters, owners, and operators of facilities that treat, store or dispose of hazardous waste must adhere to a series of regulations and permitting requirements.

Within the US waste is viewed as any other commodity and restrictions in interstate movements are prohibited. Confirmed by the US Supreme Court time and time again, states cannot restrict or ban inflows of either solid or hazardous wastes. The most recent challenge came from Alabama. The state first charged a US $72 per ton extra fee on out-of-state generators (Kemezis 1991a). Then in 1989 it banned hazardous-waste shipments from 22 states without hazardous-waste disposal facilities. The blacklist and high fees were designed to reduce out-of-state access and increase the costs for these waste generators in the hope they would go elsewhere. By its refusal to hear the case, the Supreme Court reaffirmed that wastes (hazardous and non-hazardous) are interstate commodities and thereby are constitutionally guaranteed freedom of movement among states. Higher fees for out-of-state generators are legally viewed as discriminatory, and have also been struck down by the courts. Ultimately, Alabama was forced to accept the out-of-state hazardous waste as well as a reduced fee for doing so (Kemezis 1992).

The result is a very uneven distribution of waste generators and waste recipients within the United States. Figure 6.12 illustrates the import/export trade in hazardous waste between states. The exporting states are concentrated in New England and the Great Plains. Tennessee is the largest hazardous-waste generator based on RCRA reporting (more than 13,932 lbs per capita), yet only 384 lbs per capita remains in the state (Hall and Kerr 1991). Tennessee has the dubious honour of being the largest net exporter of hazardous waste (13,548 lbs per person) followed by Nebraska (672 lbs per capita) and New York (647 lbs per capita). On the import side, Utah is the country's largest net importer of hazardous waste, with 4431 lbs per capita. This figure is bound to increase over the next few years as two new hazardous-waste incinerators began operating in 1992. These incinerators are the first in the West and handle about 80% of the out-of-state waste coming into Utah, charging haulers US $35 per ton (*versus* US $10 per ton for in-state) (Kemezis 1991b, 1992).

In terms of risk sharing, that is the percentage of waste generated that strays in the state, Arkansas and North Dakota bear a disproportionate share of the burden of other states' hazardous wastes compared to their own. Both receive more than ten times the amount of waste they generate. Many states, including some heavily industrialized ones (New Jersey, Pennsylvania, West Virginia), keep between 95 and 100% of their hazardous waste within the state.

International routes

Not only does hazardous waste flow between states, the US is also an exporter of hazardous waste as well, most of it to Canada. In 1987, for example, the US exported 75–90% of its hazardous waste to Canada (127 million tons) (Uva and Bloom 1989). A year later, the US EPA

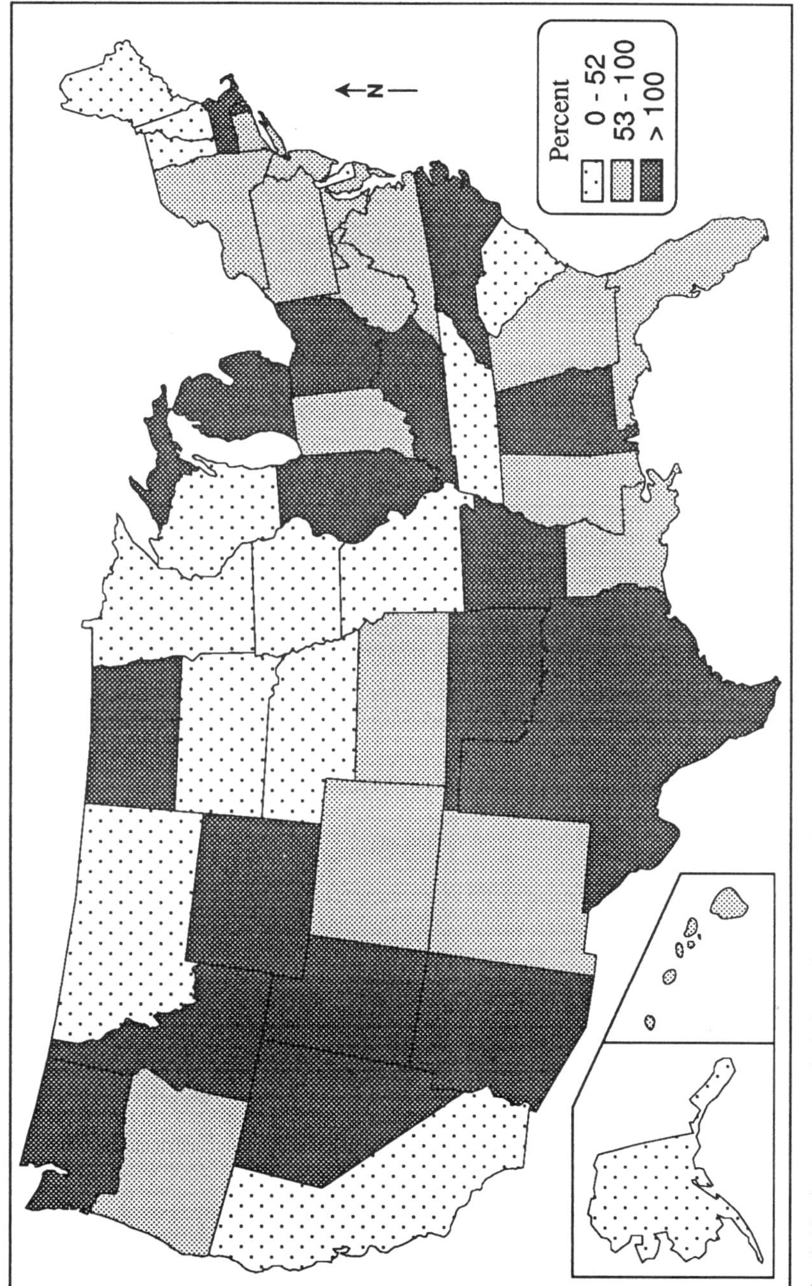

Fig. 6.12 Interstate trade in hazardous waste. The percentage of waste generated in the state that remains there. States with 50% or less are labelled exporting, while those states with more than 100% are net importers. Data from Hall and Kerr 1991.

approved the export of 3.7 million tons to 13 other countries. The European Community's record is no better. Generating three times the amount of hazardous waste as it has the capacity to handle, the only alternative is to export some of it. For years, the EC dumped on central and eastern Europe, especially Poland and the former East Germany. As part of the reunification process, Germany must now contend with decades of dumping on its eastern sister. Since 1986 the US and Europe have shipped hazardous waste to many developing countries including Brazil, Haiti, Mexico, Nigeria, Lebanon, Syria, Venezuela, Zimbabwe, and Syria. Other countries have been approached for waste trade schemes especially African countries.

Stemming the trade

The problem of transboundary shipments has become so acute that many nations have called for a ban on hazardous-waste imports in an attempt to stem the toxic terrorism inflicted on poorer nations by industrialized ones. In response to the hazardous-waste scandals in 1988 (the *Karin B.* and *Khian Sea*), the United Nations convened a Convention on the Control of Transboundary Movements of Hazardous Wastes and their Disposal held in Basel, Switzerland, in March 1989. This international treaty, signed by 54 nations including the US and most of the industrialized countries, restricts and controls the international traffic in hazardous wastes, but does not ban it outright. In addition to regulating the transboundary movements of hazardous waste (Table 6.3), the Basel Convention made the following provisions:

1) Prohibits the export of non-hazardous solid waste, hazardous waste, ash from solid-waste incinerators, and infectious waste for disposal unless the receiving country guarantees the waste will be managed in an environmentally sound manner.
2) Prohibits all waste shipments to Antarctica.
3) Exporting countries must notify importing countries in advance of any waste shipments. Exporting countries must also receive written consent from officials in the importing country before the wastes can be moved.
4) Exporting countries have the responsibility to ban shipments to consenting importing countries if the exporting country has reason to believe that the wastes will not be managed in a safe manner.
5) Prohibits all shipments from a country that has signed the treaty to one that has not.
6) The export country assumes full responsibility for proper removal and disposal of wastes shipped illegally if the waste broker cannot be identified. The convention does not require exporting countries to assume liability for any associated clean-up from environmental contamination resulting from disposal.

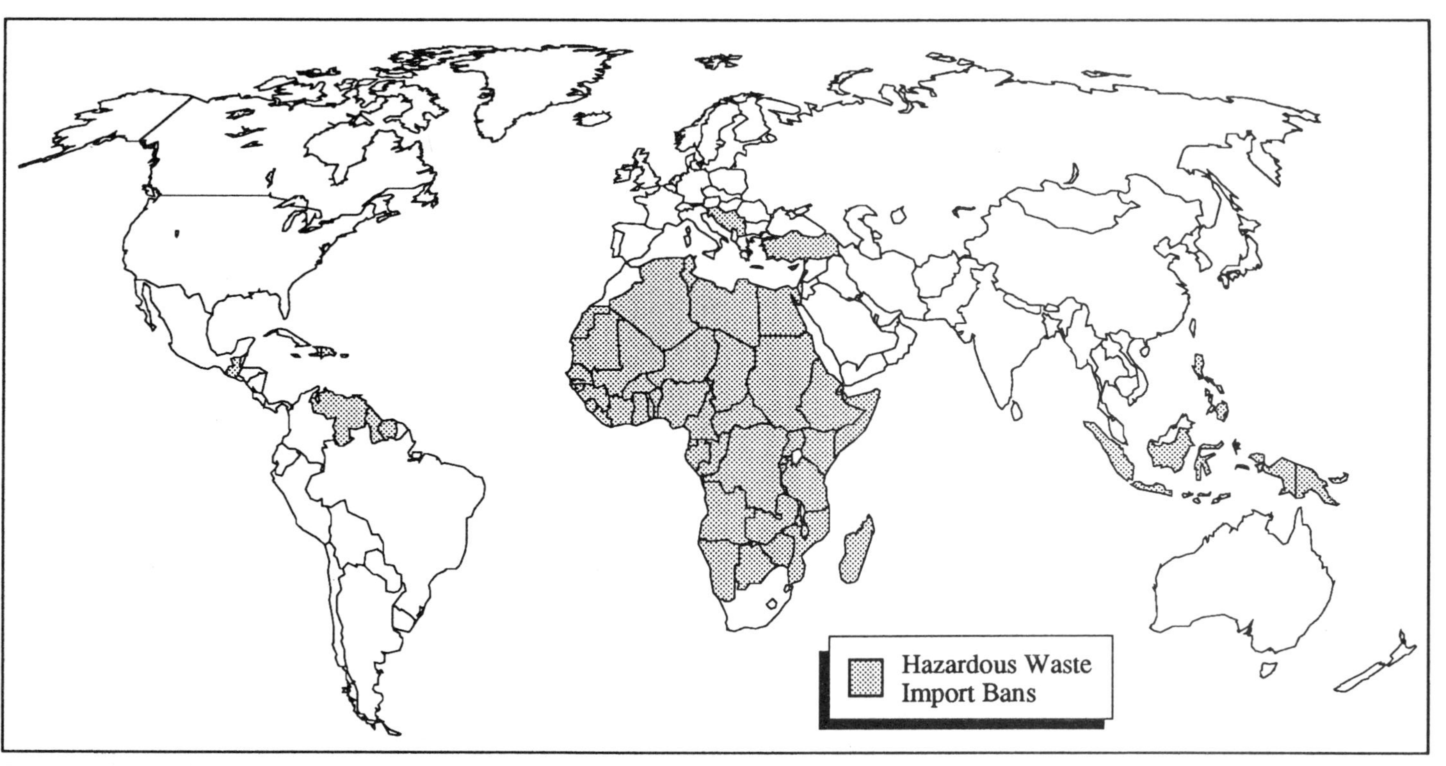

Fig. 6.13 Bans on importing hazardous waste. Most of the less industrialized countries now ban (national law or regional agreement) the import or transboundary movement of hazardous waste. Data from Greenpeace 1990.

African countries refused to sign the treaty and continue to do so, demanding a total ban on waste exports rather than mere 'informed consent' regulation. In 1989 68 developing countries – from Africa, the Caribbean and the Pacific (called the ACP) – and the European Community negotiated an agreement called the Lome Convention which bans all hazardous waste and radioactive waste shipments from EC countries to the ACP countries. In addition, the ACP countries agree not to import hazardous or radioactive waste from any other non-EC country (Greenpeace 1990).

More recently, the Organization for African Unity (representing every African country but South Africa and Morocco) called for a complete ban on hazardous-waste imports. In 1991 the Bamako Convention was signed by 17 African countries. This regional agreement bans the import of hazardous waste into Africa and controls the transboundary movements of hazardous waste within Africa as well. Most of the world's developing countries now have national laws or regional treaties that completely ban hazardous-waste imports (Fig. 6.13), taking a much stronger stance than the Basel Convention.

Who's at risk?

It is clear from the preceding discussion that the risks associated with the production, use, and disposal of hazardous materials are pervasive, not only at the local and national levels, but internationally as well. Much media attention has focused on the international trade in hazardous waste and the American and European role in it. They are, after all, the largest hazardous-waste generators in the world. There is a clear pattern of global dumping on developing countries from industrialized nations who are struggling with the rising amounts of hazardous wastes and rocketing costs of disposal. This situation, coupled with dire economic conditions among Third World, eastern, and central European (including the CIS) nations, and non-reputable waste brokers sets the stage for the international trade in hazardous waste. Despite international efforts, there continues to be inequities in risk burdens as poorer states and nations become the hazardous-waste dumping grounds for wealthier nations, be it in Selma, Alabama, Sulina, Romania or Schoenberg, Germany. Nor has there been any relief from the risk burdens imposed on millions of people in the industrialized nations from decades of hazardous-waste production and illegal disposal right in their own backyards.

7

Fire in the rain

'Wormwood is a bitter herb, used traditionally by country folk as a spring tonic. In Ukrainian its name is . . . chernobyl' (Gould 1991, p. 24).

One of the most perplexing and politicized risks that confront us is nuclear technology. Whether used for peaceful purposes or in military weapons, the atom symbolizes the nuclear age and the wide range of hazards associated with it. This chapter examines nuclear hazards, not just those from commercial nuclear power plants, but also those risks from the nuclear weapons complex. Needless to say, this chapter is not an exhaustive treatment of the subject; many books have been written on various aspects of the 'nuclear dilemma'. Rather, I hope to illustrate the divergence between technocratic and public views of risk and public unwillingness to accept nuclear hazards. In so doing, I will try to weave together the stories of the commercial nuclear industry and its governmental-sanctioned counterpart, the military.

Atoms for peace?

The nuclear age was ushered in that fateful day in 1941 when Enrico Fermi achieved the first controlled nuclear chain reaction under the west stands of Stagg Field, the football stadium of the University of Chicago. Today the site is home to the University's main research library and is commemorated by a small sculpture. The site is also on the city tour of Chicago and is frequented by many tourists, especially Japanese visitors.

The race for nuclear fission had started two years earlier when experimental evidence of the fission process (the splitting of the uranium atom into two thereby releasing energy) was discovered by two German chemists and reported in the international scientific journals. Seeing the potential for weapons of mass destruction as well as a potential new source of power, physicists in the US (many of whom were recent immigrants from

Europe fleeing Hitler's repression) sent a letter to President Roosevelt advising him of the possibility of the atom bomb. This was in 1939.

It was not until early 1941, however, that Roosevelt took the warning seriously and committed resources under the aegis of the US Army Corps of Engineers to build an atomic bomb before the Germans. Roosevelt's actions resulted from two considerations: (1) the 1940 discovery of plutonium by Glenn Seaborg and others at Berkeley, which is an artificial element that sustains the fission process better than uranium; and (2) a report by the British military that, in fact, an atom bomb was technically possible (Titus 1986). The Manhattan Project was thus born on 13 August 1942, and the rest, as the saying goes, was history. What is interesting is that up until 1942, there was a free exchange of ideas, hypotheses, and experiments in nuclear physics among the international scientific community, including the Germans. With the establishment of the Manhattan Project, the US government-sponsored nuclear research effort became cloaked in secrecy, and remains that way even today.

With the delivery of the atom bomb and a successful test during war (6 August 1945 in Hiroshima, and 9 August 1945 in Nagasaki), the US became the purveyor of nuclear technology: its advocate, guardian, and controller. To combat the horrors that the American public witnessed on the military use of the technology and to preserve the military, industrial, and research infrastructure built during the Manhattan Project years, President Truman asked Congress to create a post-war commission to control the use and production of atomic power. The Atomic Energy Act of 1946 did just that by establishing the Atomic Energy Commission (AEC). Nuclear development was theoretically placed in the hands of civilians, not the military, yet the military continued to dominate policy and thus the direction of the AEC. Because such a large portion of the AEC's work involved some aspect of weapons development, the military was inextricably linked and continued to keep nuclear technology away from public scrutiny.

The military monopolized nuclear technology until 1953 when President Eisenhower delivered his 'Atoms for Peace' speech before the United Nations on 8 December 1953. Calling for international cooperation in the development of peaceful applications of atomic energy (including electricity generation), this programme stimulated private-industry interest in developing nuclear power as a viable energy source. With the passage of the Atomic Energy Act of 1954 (which amended the 1946 law), the governmental monopoly was supposedly ended setting the stage for privately owned, but governmentally regulated nuclear facilities.

Almost from the beginning, controversies were rampant in the nuclear industry. The AEC concentrated its efforts on research and development programmes to build a prototype commercial reactor with the help of electrical utilities and reactor manufacturers—groups that the agency also had to regulate. Early warnings were heard about the unsafe technology, the problem of waste disposal and how the regulation of the industry was ineffectual since the AEC was both the promoter of nuclear technology as well as its regulator (Ford 1982). By law and in practice, the AEC was an inherently conflicted agency. The military continued to control weapons development, while the civilian divisions were involved in research, development, and regulation of atomic energy programmes.

In 1974 Congress tried to rectify the problem of inconsistent agency missions by passing the Energy Reorganization Act. Some suggest that it was also responding to the nuclear industry which was overly critical of the costly and time-consuming licensing process and regulations (Quirk and Terasawa 1981). The law abolished the AEC and established two new agencies with differing missions. The Energy Research and Development Agency (ERDA) was established to direct federal research and development into all energy sources, including nuclear. The Nuclear Regulatory Commission's (NRC) mission was to license and regulate the private nuclear power industry. Three years later (1977) Congress established a new cabinet-level agency, the Department of Energy (DoE), thus consolidating various energy-related activities and programmes including ERDA, into one cabinet department. The DoE has both civilian and military research and development activities. The bifurcation of American nuclear policy into military *versus* commercial nuclear programmes, one shrouded in complete secrecy, the other more open to public scrutiny and debate, has had a direct bearing on the geography of nuclear hazards.

Ionizing radiation

Radioactivity or more precisely ionizing radiation occurs naturally, e.g. in the Earth's crust, and from cosmic rays entering the Earth's atmosphere. Our concern with ionizing radiation is its ability to cause widespread biological damage especially among humans. Radiation exposures result in both direct and indirect effects. The direct impacts called 'somatic effects' become evident almost immediately (days to weeks) in the case of large exposures, or within years from lower levels of exposure. Leukemia and skin cancer, radiation sickness, radiation burns, and cataracts, are all examples of somatic effects. Indirect impacts or 'genetic effects' are not seen in the exposed individual but are passed on to future generations through chromosomal changes and mutations. The hereditary material can impact generational fertility and produce congenital deformations. The somatic effects of ionizing radiation exposure became known a short time after Roentgen's discovery of X-rays in 1895. The genetic effects were recognized much later as were the indirect pathways that exposures took, such as inhalation and ingestion of contaminated food, water or air.

When we speak of risks of nuclear hazards, we are referring to the additional doses of radioactivity from human activities that are beyond the 'normal background levels'. These include the release of large amounts of deadly fission from reactor accidents or radioactive waste products from nuclear activities. High-level wastes (HLW) are waste products from reactors, their spent or used fuel. The radioactivity has a half-life (the amount of time it takes for half the atoms to disintegrate) of more than 1000 years. Intermediate-level waste (ILW) has a shorter half-life, but there is more of it. One example of intermediate waste is the metal cladding around the spent fuel rods. The fuel can be separated out and reprocessed, the metal casings become waste products. Low-level waste (LLW) is also generated by the nuclear industry. However, medical facilities, universities, and some industries also produce low-level waste such as contaminated protective clothing. The difficulty in assessing nuclear risk depends on the type of radioactivity released, the pathway of exposure (direct, inhalation, ingestion), and the length of time it takes for the radioactivity to decay or its half-life.

Two terms are important in understanding these risks. The first is the 'becquerel' (Bq) which measures the quantity of radiation from a given mass of material; the amount of radioactivity released from a nuclear power plant accident, for example. Previously, the term 'curie' was used to measure this. The second term, the 'sievert' (Sv), is the amount of radiation absorbed by the human body and measures the level of human exposure to ionizing radiation. Because of the direct link between exposure and biological damage (dose and response), radioactive exposures (both occupational and non-occupational) are tightly monitored for all users of these materials (medicinal, commercial, military).

Strict standards are set for controllable exposures. In 1952 the International Commission on Radiological Protection (ICRP) set the general public exposure limit at 15 millisieverts (mSv). By 1959 this dose was lowered to 5 mSv, where it stands today (Mounfield 1991). To gauge the impact of radiation exposures, Table 7.1 provides a sampling of exposures with their known effects. We have very detailed knowledge on the effects of massive doses of radiation as you can see. There is more scientific controversy over the physiological responses to low-level doses, especially their genetic effects. The determination of safety levels has been fought in the halls of science and by regulators, yet the acceptability of such estimates is very much in the public domain. Thus, the stage is set for very radical interpretations of the risks of nuclear technology and its continued and future use.

The geography of nuclear hazards

Space and time are the defining characteristics of nuclear hazards—they are geographically disparate and last from minutes to millennia. The hazards begin with the mining of uranium and continue through the

Table 7.1 Radiation exposures.

Dose	Impacts
100 Sv	Death within days, central nervous system damage
10–50 Sv	Death in 1–2 weeks, gastro-intestinal damage
3–5 Sv	50% of the exposed group dies within 1–2 months, damage to bone marrow
2 Sv	Skin burns, bone-marrow damage, some survivors
0.1 Sv (100 mSv)	Temporary blood-count depression, temporary sterility, mental and growth retardation in children
0.005 Sv (5 mSv)	General public exposure standard from the ICRP
0.001–0.003 Sv (1–3 mSv)	Level of naturally occurring radiation

Sources: Mounfield 1991; May 1989.

fabrication of fuel or weapons, the production of electricity, and the radioactive waste that remains often hazardous for millennia. Throughout each phase in the cycle, materials must be transported from place to place thereby creating both stationary and mobile sources of hazards (Fig. 7.1).

The nuclear cycle

Most of the activities in the nuclear cycle are geographically dispersed. The mining and milling of uranium ore in the US, for example, occurs mostly in the West, especially in New Mexico and Wyoming. Since the naturally occurring uranium cannot sustain the fission process it must be enriched to increase the concentration of fissionable material. At this point, the nuclear cycle branches out depending on the end use, e.g. nuclear fuel for the navy, components for weapons manufacture, and commercial reactor fuels (Fig. 7.1). The conversion process for commercial fuel rods is carried out in four facilities in the US (Illinois, Oklahoma, Pennsylvania, and South Carolina). The enrichment process and the fabrication of the material into fuel rods occurs in only three facilities (Oak Ridge, Tennessee, Paducah, Kentucky, and Portsmouth, Ohio). Once fabricated, the fuel rods are sent to commercial reactors throughout the country. On the military side, uranium processing and conversion occurs at the Feed Materials Production Center in Fernald, Ohio, and the Idaho National Engineering Laboratory, while nuclear materials production and processing (plutonium and tritium) is done at Hanford, Washington, and Savannah River, South Carolina. Warhead-component production occurs at plants in Colorado, Tennessee, Ohio, Florida, and Missouri. The final

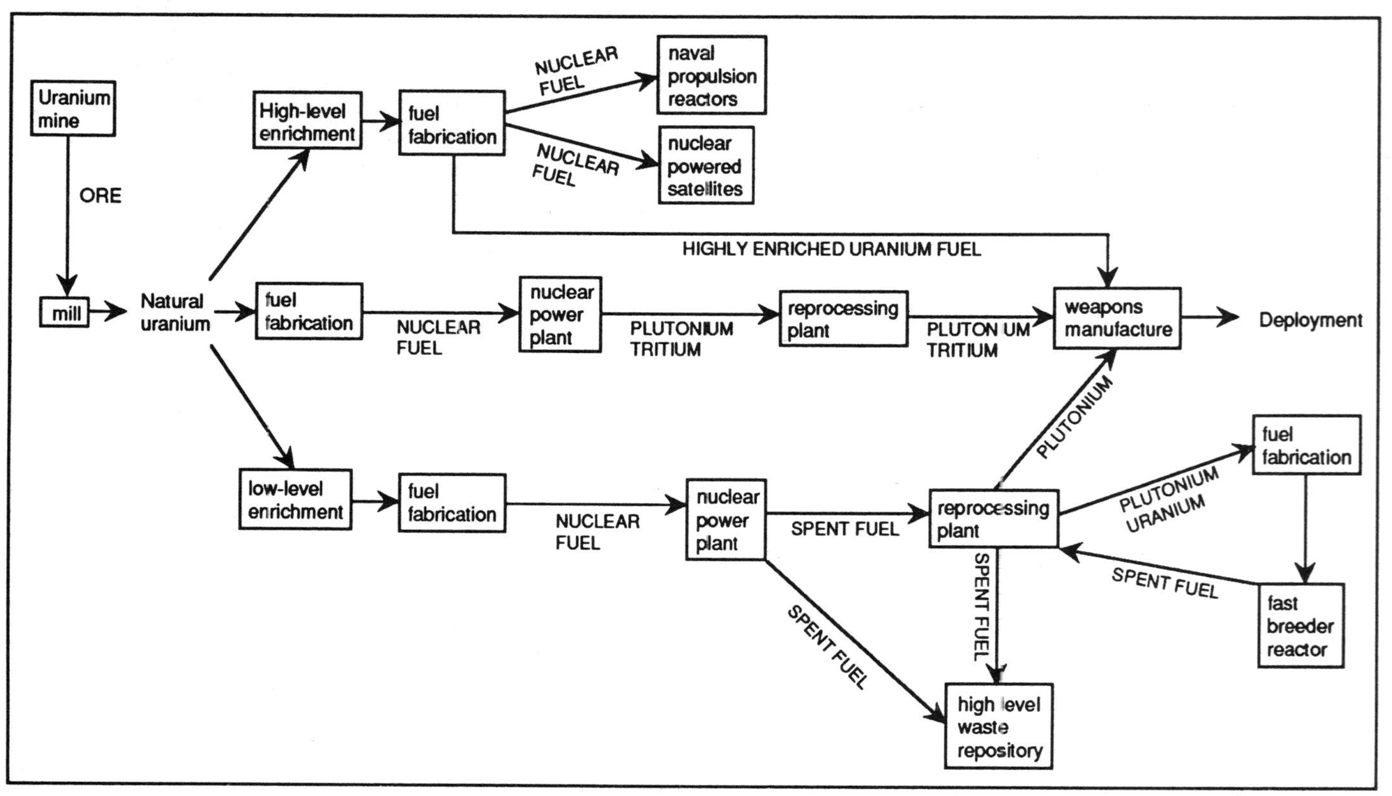

Fig. 7.1 The nuclear cycle. Nuclear hazards are created from military and civilian uses of fissionable material, from the mining of the uranium to the disposal of waste and from stationary facilities and transportation networks.

assembly of nuclear weapons is done at the Pantex Plant in Amarillo, Texas (see Fig. 6.10 for a map of the DoE weapons complex). The weapons are then distributed to US military forces worldwide. Spent fuel from reactors is sent to a reprocessing facility (currently there are none operating in the US) to recover some of the fissionable material to be reused as fuel. The last stage in the nuclear cycle is the storage of radioactive waste. There is no permanent high-level waste repository in the US at this time.

At the international level, the nuclear cycle is also geographically dispersed. In addition to the US the primary producers of uranium ore include Canada, South Africa, France, Germany, Namibia, Niger, Australia, Czechoslovakia, China, and the Commonwealth of Independent States (CIS). In an attempt to gain more foreign currency, the former Soviet republics of Kazakhstan, Kyrgyzstan, Russia, Tajikistan, Ukraine, and Uzbekistan have exported most of their uranium production since 1990. The US is a large importer of this cheap uranium for its commercial nuclear reactors, since, by law, only US-mined uranium can be manufactured into weapons. Enrichment facilities are concentrated in 12 nations (Argentina, Brazil, China, France, Germany, Japan, Netherlands, Pakistan, South Africa, United Kingdom, US, and Russia) Refining, conversion, and fabrication of nuclear materials into fuel rods is heavily concentrated in Canada, France, UK, CIS, and the US.

Throughout each of these stages, we have to remember that the materials are transported either by rail or truck (continental scale) or ship (trans-oceanic), placing untold millions of people at risk should there be an accident resulting in spillage of these materials (Jacob and Kirby 1990). The stationary hazards at each stage are also worth noting. They range from radioactive-mine tailings, low- and high-level waste from the reactors, discharges into the environment (vented steam, slightly radioactive water), and of course accidents (Bartimus and McCartney 1991). It is also important to note that nuclear weapons are deployed around the globe (on the land, in the sea, and in the air), providing both stationary and mobile sources of hazards as well.

Nuclear America

To map all the places where nuclear hazards are produced is a daunting task, especially if you include both the military and civilian producers. What we tend to see more often is the geographic depiction of commercial reactors, one indicator of nuclear hazards (Fig. 7.2). As of August 1991 there were 119 operating reactors in the United States at 68 different sites. Most are concentrated in the East, where electrical demand is high and sources of coolant (water) are more plentiful.

However, there are some recent attempts to depict the extent of the US nuclear arsenal both at home and abroad (Arkin and Fieldhouse 1985).

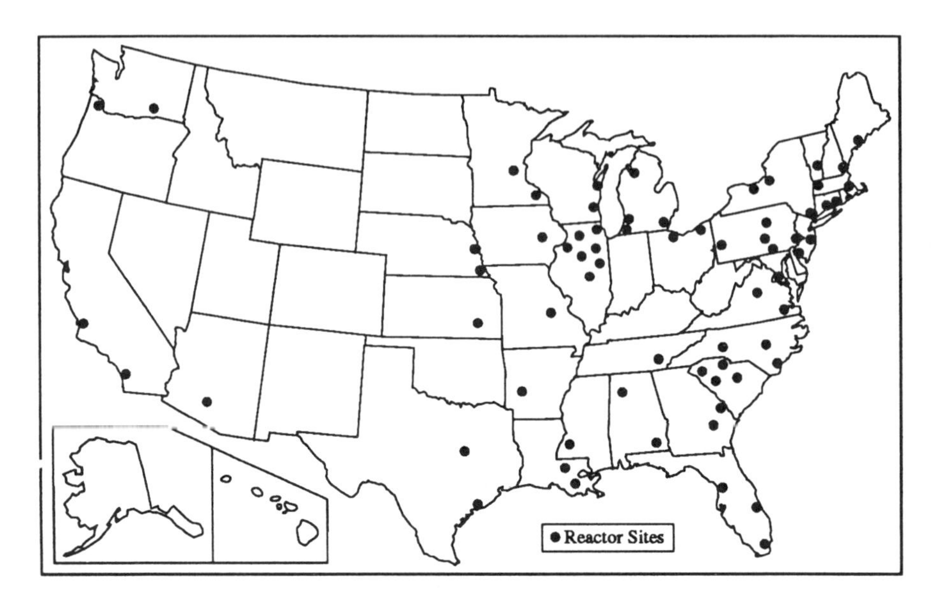

Fig. 7.2 Location of US commercial reactors, 1991. These are 70 sites with operating commercial nuclear power plants, many with more than one operating reactor. Source: Nuclear News 1991.

Within the US the spatial distribution of nuclear arsenals is concentrated in the Great Plains and Pacific Coast regions (Fig. 7.3). In 1985 states with the largest number of nuclear-warhead deployments were South Carolina (1962), New York (1900), North Dakota (1510), and California (1437). In terms of total military facilities, North Dakota, California, and Montana rank as the top three. Finally, California, Alaska, and Maryland lead the nation in the number of nuclear facilities including research and development centres with 79, 42, and 35 respectively (Arkin and Fieldhouse 1985). The rest of the weapons complex, depicted as contaminated processing plants, laboratories, etc., is mapped as well (Fig. 7.3). The largest concentrations of these facilities are found in the Buffalo, New York, metropolitan region. California is the most nuclearized state in the Union. The recently announced military-base closings and clean-up programmes will alter the map of America's nuclear hazards during the 1990s and beyond (Schneider 1991a).

Worldwide nuclear hazards

It is even more difficult to map worldwide nuclear hazards because of the sensitivity of nuclear weapons programmes and deployment to national security. We know, for example, that there are 11 members of the prestigious nuclear nations club; nations that possess the scientific and technical capability to manufacture an atomic bomb. The US,

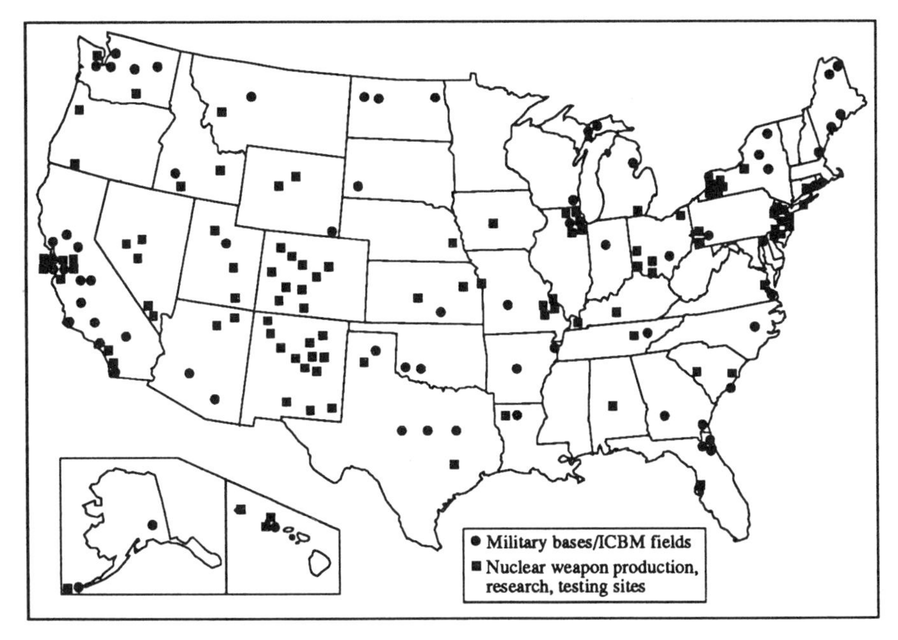

Fig. 7.3 US nuclear battlefields, 1990. Locations of nuclear hazards from the military use of nuclear technology. Hazards are found in military bases, Intercontinental Ballistic Missile launching fields, nuclear weapons research and production complexes, and weapons testing sites. Sources: Arkin and Fieldhouse 1985; Schneider 1991b.

CIS, United Kingdom, France, and China were founding members. India, Israel, Pakistan, Argentina, Brazil, and South Africa converted peaceful technology (power production) into weapons capability, despite technological controls placed on them by the founding members. Iran, Iraq, North Korea, and Libya are within grasp of the atomic prize in just a few years. Some argue they already have it.

Despite the dearth of information on international nuclear hazards, we can make some educated guesses on their geographical extent. The first is to map the patterns of weapons production and deployment as much as we know it. The other alternative is to examine the distribution of commercial reactors.

There are 418 operating reactors worldwide as of August 1991, including the 119 in the United States (Table 7.2) (Nuclear News 1991). Unlike the relatively dispersed pattern in the US, many nations prefer to concentrate the number of reactors at one site, such as the eight reactors in Pickering, Ontario, Canada, or the six in Ohkuma, Fukushima prefecture in Japan, or Gravelines, Nord, France (Fig. 7.4). In so doing, the number of sites is reduced and so presumably is the spatial extent of the hazard. In Canada, for example, there are 19 operating reactors, but they are concentrated in just five locations: Bay of Fundy, New Brunswick (one), Pickering, Ontario, just outside Toronto (eight), Tiverton, Ontario, along

Table 7.2 Number of operating commercial reactors, 1991.

Country	No. sites	No. reactors
Argentina	2	2
Belgium	2	7
Brazil	1	1
Bulgaria	1	5
Canada	5	19
Commonwealth of Independent States	15	42
Czechoslovakia	2	8
Finland	2	4
France	19	55
Germany	16	21
Hungary	1	4
India	4	7
Japan	17	41
Mexico	1	1
Netherlands	2	2
South Africa	1	2
South Korea	4	9
Spain	7	9
Sweden	4	12
Switzerland	4	5
United Kingdom	14	37
United States	70	119

Source: Nuclear News 1991.

Lake Huron (eight), Newcastle Township, Ontario (one), and Becancour, Quebec (one). While there is little, if any, scientific proof on the relation between concentrated *versus* dispersed reactors and their probability of failure, it does provide some food for thought. The spatial distribution of operating reactors does provide some insight on the location of nuclear hazards.

On the military side, nuclear forces dot the landscapes in the former Soviet Union, France, United Kingdom, and China (Figs 7.5 and 7.6). In the former Soviet Union, the nuclear arsenals were concentrated in the western portion of the country along the Trans-Siberian railroad, and in the Arctic staging areas (Fig. 7.5a). Recent reports claim that more than 900,000 people in the CIS formerly engaged in the nuclear weapons complex. Many of these people lived in top-secret cities, which simply disappeared from the map during the Cold War decades. Some of these sites (design laboratories, warhead-assembly locales, and uranium-enrichment facilities) have just been made public, some that the US did not know about (Broad 1992). Others they did such as the centres for research and weapons design at the Arzamas laboratory south of

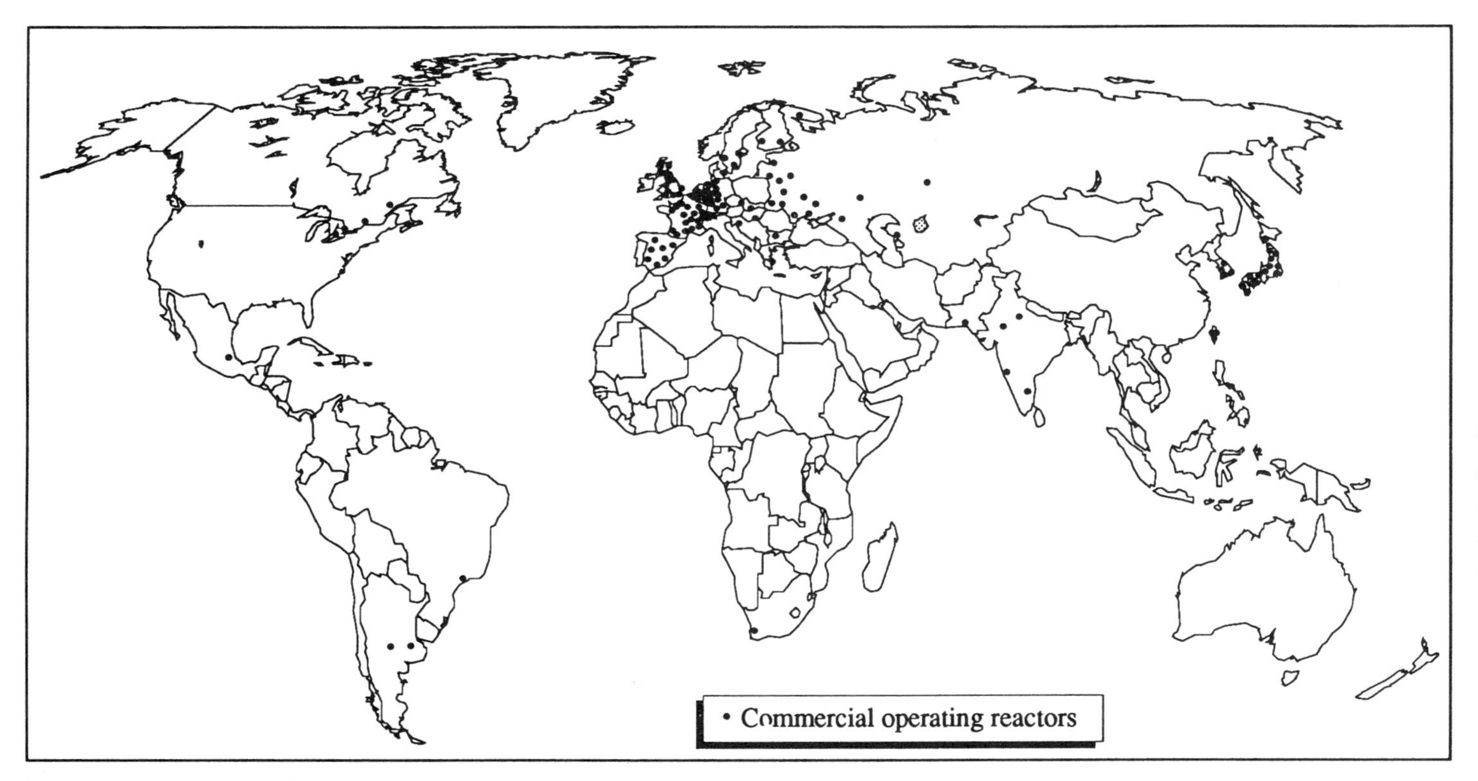

Fig. 7.4 Worldwide operating commercial reactors, 1991. Reactors are concentrated in the industrialized countries and are clustered into a relatively few sites with multiple reactors at each site, quite a different pattern than in the US (see Fig. 7.2). Source: Nuclear News 1991.

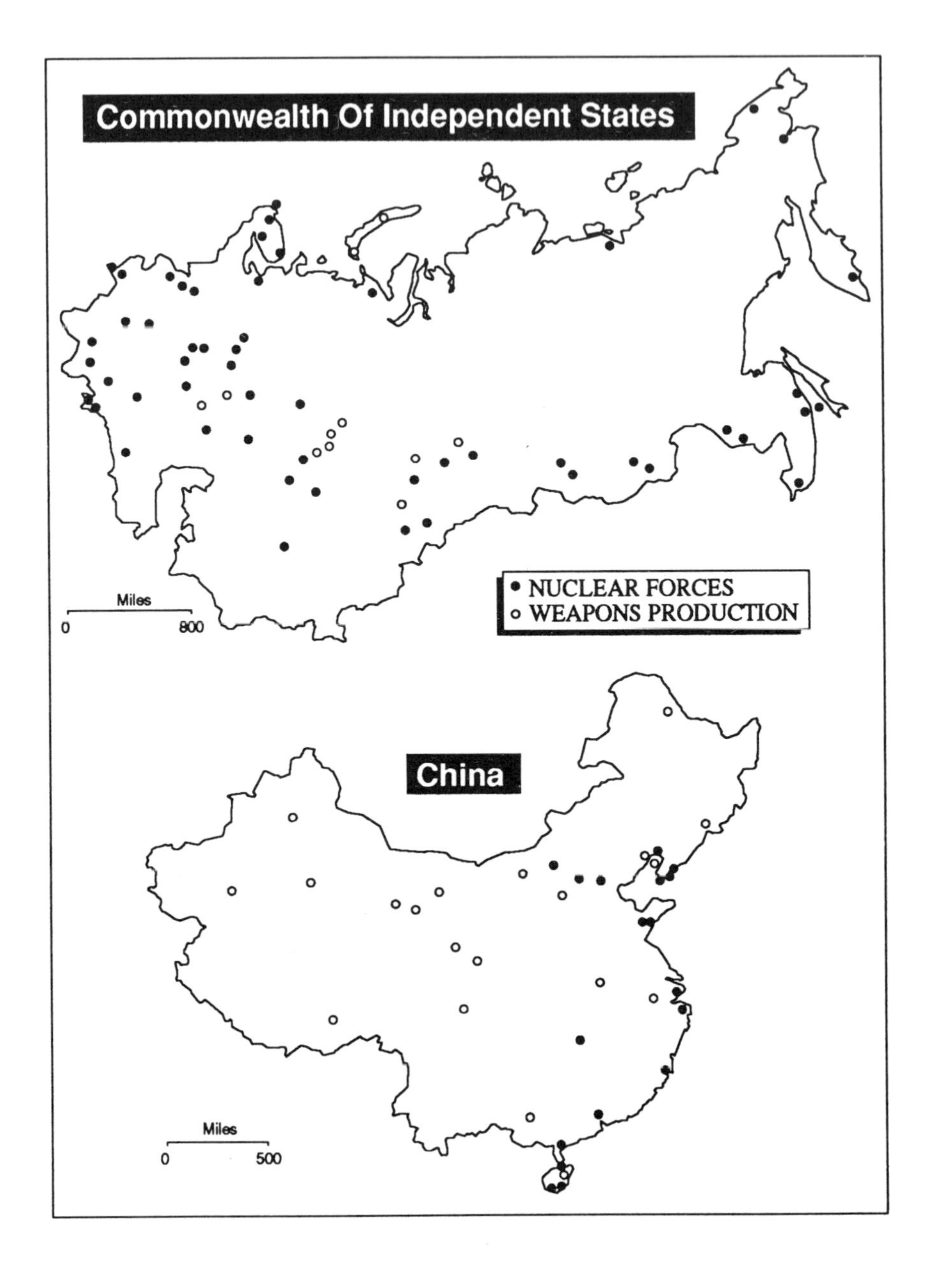

Fig. 7.5 Soviet and Chinese nuclear forces, 1985. The nuclear forces of (a) the former Soviet Union and (b) China are difficult to map accurately because the deployments often move to avoid detection. Source: Arkin and Fieldhouse 1985.

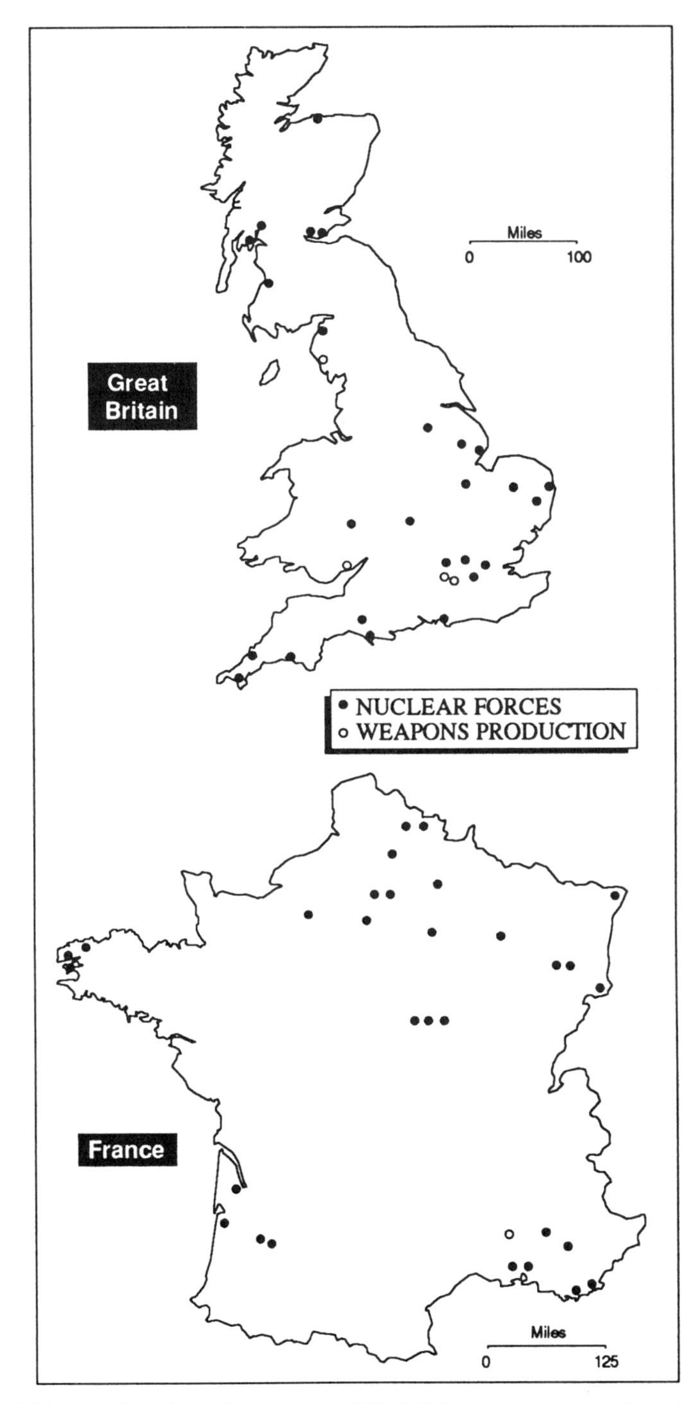

Fig. 7.6 British and French nuclear forces, 1985. British weapons production (a) is scattered around the country while nuclearized military bases are concentrated in the south and southeast. French weapons production (b) is located in the southern part of the country. Source: Arkin and Fieldhouse 1985.

Nizhny Novgorod (Gorky), and at facilities in Kyshtym, Moscow, St. Petersburg (Leningrad), and Semipalatinsk. Production plants are located in Yakaterinburg (Sverdlovsk), Novosibirsk, and Chelyabinsk. Two testing sites (out of as many as 20) are still active: the island of Novaya Zemlya and Semipalatinsk in Kazakhstan.

China, the most unknown of the nuclear powers, has most of its arsenal massed along the northern frontier region near its border with Russia (Fig. 7.5b). Nuclear materials are produced at Lanzhou, Yumen, Baotou, Hong Yuan, Jiugquan and Urumqui. Weapons are produced and assembled in three locations—Lanzhou, Boatou and Haiyan—that are known. Weapons testing takes place at Lop Nor in western China.

The UK's forces are concentrated in the southeast portion of the country (Fig. 7.6a). The map is somewhat deceiving, however, because it does not include the US military's storage of nuclear weapons in Britain; a quantity greater than the entire British arsenal. The British nuclear weapons complex employs 8000 people. Research occurs primarily at Aldermaston in Berkshire which also has a production facility. Nuclear materials are produced at Calder Hall and Windscale (BNFL's Sellafield Site) in Cumbria, and Chapelcross in Dumfries and Galloway. Weapons are produced at the Royal Ordnance factories in Cardiff and in Burghfield, Berkshire, where the final warheads are assembled. Since 1962 British warheads have been tested at the Nevada Test Site in the US.

France's nuclear arsenals are located in the northwest border region as well as along its coasts (Fig. 7.6b). Nuclear materials for French nuclear weapons are produced at Marcoule, Miramas, and Pierrelatte in southern France. Quite a few chemical accidents have occurred in Pierrelatte during the last 30 years (see Chapter 5). Little else is known about the French weapons complex other than their current test site is the Mururoa Atoll in French Polynesia.

In addition to domestic bases, both the US and the former USSR exported their nuclear arsenals under the guise of strategic defense. Europe was the most targeted area, having the greatest concentration of nuclear weapons anywhere in the world. Prior to 1990 the nuclear arsenals were clustered in three areas: West Germany, East Germany, and in southeastern England (Fig. 7.7). There were 150 primary NATO nuclear units and more than 30 primary Warsaw Pact units (encompassing more than 4000 nuclear warheads) in East and West Germany alone (Arkin and Fieldhouse 1985). The Pacific Rim also has nuclear arsenals, some on land. Vietnam and the Pacific region of Russia house naval-support facilities, especially in Vladivostok, and Petropavlovsk (Young 1983). US nuclear arsenals are placed in South Korea (storage sites for warheads, and airbases) and Guam, which houses the main stockpile of nuclear weapons in the Pacific.

In the early 1990s, there were about 3000 warheads located in Third World regions (Arkin and Fieldhouse 1985). Naval weapons are the most

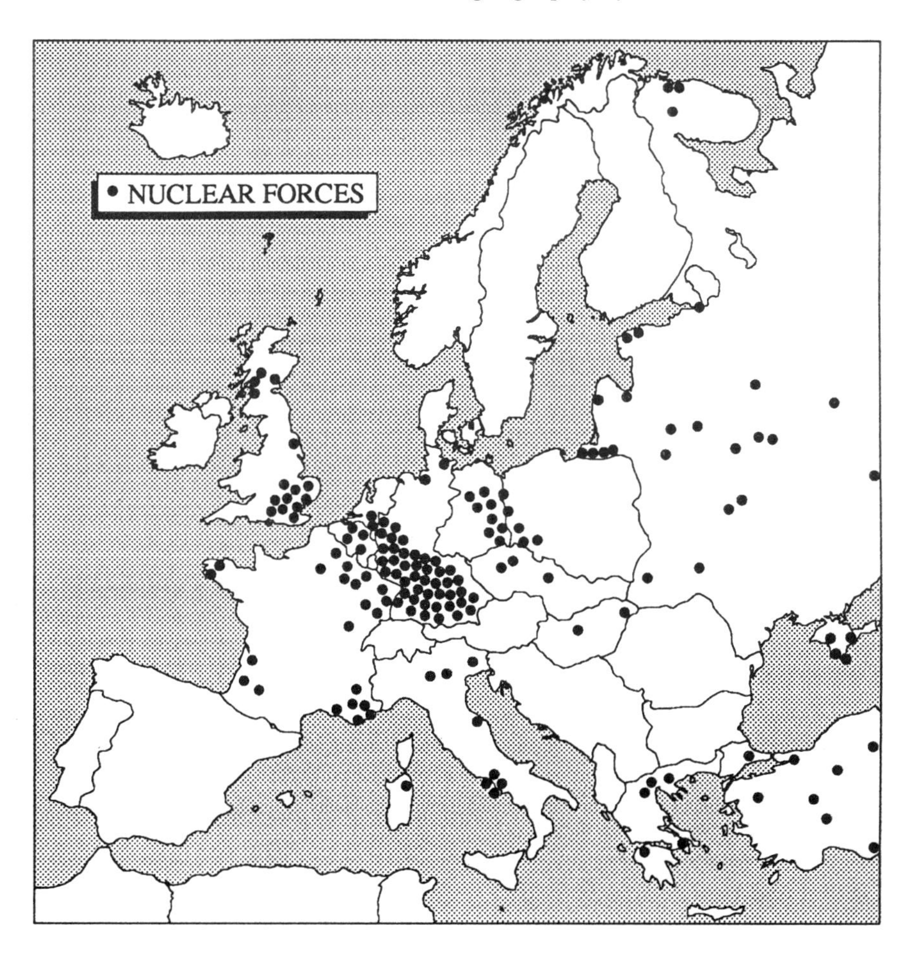

Fig. 7.7 European nuclear battlefields, 1985. Central Europe had the largest concentration of nuclear weapons in the world. With the demise of the former Soviet Union, and the democratization throughout Europe, the nuclear forces have been severely reduced. Source: Arkin and Fieldhouse 1985.

numerous, particularly aircraft carriers and submarines. Soviet, US, British, and French naval forces routinely patrol the eastern Mediterranean, Persian Gulf, Indian Ocean, Caribbean Sea, and the South Atlantic Ocean. On land, Turkey hosts NATO and US nuclear forces including a nuclear weapons storage site at Incirlik.

The export of nuclear arsenals by the US and the former USSR had a distinct geographic pattern. US nuclear exports were in the form of strategic naval forces in the Indian Ocean, Pacific Ocean, and Mediterranean Sea, as well as land-based support in NATO countries in Europe. Former Soviet strategic forces were concentrated on the African continent and included both naval and air-reconnaissance facilities. With the dismantling of the former Soviet Union, the geography of nuclear hazards has changed dramatically in just a few short years. In 1985 there were between 49,000 and 59,000 warheads in the world's nuclear arsenals; 97% of them owned by the US and the former Soviet Union. By 1991 less than 22,500 warheads were part of the US and Russian arsenals, a reduction by more than a third (Friedman 1992). Nuclear arsenals are no longer present in central and eastern Europe, having been recalled by the former Soviet Union. US forces are also retrenching, but not as quickly. The tentacles of US nuclear arsenals are still felt worldwide.

Accidents happen

Accidents have always been part and parcel of the nuclear technology. Harry Daghlian was a member of the Manhattan Project and became the first known North American to die of radiation sickness in 1945. The second was Louis Slotin, a physicist with the Manhatten Project who was inadvertently exposed to radiation while attempting to secure a beryllium core to an assembly in preparation for one of the above-ground tests in Bikini. The accident occurred in May 1946, long after the war, but it marked the beginning of the one-sided nuclear arms race, and the intensification of nuclear hazards.

Broken arrows, bent spears

Despite their rather low probability of occurrence, accidents involving nuclear materials do happen, often with disastrous results. Chernobyl immediately comes to mind as does Three Mile Island. But, there are many more instances of accidental releases of radioactivity both from commercial reactors as well as from weapons and weapons manufacturing. One of the difficulties in assessing incidents is determining their severity. In 1990 the International Nuclear Event Scale was devised to communicate the significance of a nuclear accident promptly and simply to the public. The scale ranges from level 7 (a major accident such as Chernobyl) to 0 (incidents with no significance to health or the environment) (Table 7.3). The underlying logic of the scale is to distinguish accidents based on their off-site impacts, primarily the amount of radioactivity released, and the degree of damage to the reactor itself. The scale also implies certain levels of emergency-response to mitigate the effects of the radioactive exposures.

The military has its own classification of nuclear incidents that they describe with quaint euphemisms. A 'broken arrow' is an unexpected event involving a nuclear weapon where the risk of an outbreak of nuclear war does not exist (e.g. nuclear detonation, non-nuclear detonation, radioactive contamination, or theft or loss of a weapon or a component of a weapon). A 'bent spear' is also an unexpected event where a weapon

Table 7.3 International Nuclear Event Scale.

Level	Criteria
Accidents	
7 major accident	Major release, acute health effects, delayed health effects over large area, long-term environmental consequences
6 serious accident	Significant release, full implementation of emergency-response plans
5 accident with off-site risks	Limited release off-site, some selective implementation of local emergency-response plans, damage to core by mechanical effects or melting
4 accident with on-site risks	Minor release, public exposure within allowable limits, some acute health effects to workers, some core damage, may need some emergency response to protect local food sources
Incidents	
3 serious	Very small release of radioactivity above authorized limits, public exposure well within prescribed limits, contamination on-site, some worker exposure, precursor to accident
2 incident	Technical mishap that indirectly threaten plant safety, leads to re-evaluation of safety procedures
1 anomaly	No off-site or on-site risks posed, but indicative of lack of safety provisions
0 below scale	Non-significant

Source: Mounfield 1990.

is damaged or a component is damaged and must be replaced, or is an event that could lead to a nuclear weapon accident. A 'faded giant' is an uncontrolled reactor that reaches criticallity, damaging its core or releasing fission products to the environment. Finally, 'dull sword' is the loss of control of radioactive materials that present a hazard to human health, the environment, or human welfare (May 1989).

The environmental organization, Greenpeace, compiled a history of nuclear accidents listing hundreds of events since 1940. The criteria for inclusion for nuclear-reactor accidents were: off-site releases of significant amounts of radioactivity; deaths or significant injuries; core damage or severe damage to major equipment such as the emergency cooling system; the event caused inadvertent criticallity or was a precursor to a potentially more serious accident; and significant recovery costs (more than US $500,000). This study also included the military classifications described earlier. The chronicle illustrates near misses, human foibles, mechanical failures, and early warnings for greater dangers ahead. Table 7.4 provides a brief synopsis of some of the largest accidents.

Table 7.4 Major civilian and military nuclear accidents.

Date	Place	Cause/effect
2 December 1949	Hanford, Washington, USA	Radioactive plume from weapons reactor covers 200 miles by 40 miles releasing 20,000 curies xenon-133, 40 miles off-site
12 December 1952	Chalk River, Ontario, Canada	Core meltdown in experimental reactor, some off-site release of radioactivity
10 October 1957	Windscale Sellafield, UK	Fire in two plutonium reactors resulting in off-site releases of I-131, contaminated milk, core meltdown
December 1957	Chelyabinsk-40, Kyshtym, USSR	Explosion in plutonium production plant, extensive damage (635 square miles), 2 million curies of radioactivity releases, kept secret for decades
2 January 1961	Arco Idaho, USA	Reactor-core meltdown in SL-1 reactor, three killed
10 April 1963	North Atlantic Ocean, 220 miles off New England	Sinking of the atomic submarine the *USS Thresher*, killing 129 after a malfunction of its reactor, ship never recovered
5 October 1966	Laguna Beach, Michigan, USA	Partial meltdown of Fermi commercial nuclear reactor
22 March 1975	Browns Ferry, Alabama, USA	Crippling of the emergency core cooling system resulting in a near disaster, as workers checked for air leaks using a lit candle that led to the ignition of insulation around electrical system
28 March 1979	Three Mile Island Pennsylvania, USA	Mechanical failure and human error combined to cause a partial fuel meltdown
14 November 1983	Sellafield, UK	Leaking radioactive waste from discharge pipe lead to partial contamination of Irish Sea and nine-month closure of beaches in the vicinity of the plant
25 April 1986	Chernobyl, Ukraine	Worst nuclear power plant accident in history
13 September 1987	Goiania, Brazil	Abandoned medical clinic radio-active waste recovered by local

		junkyard man, killed four, 249 contaminated, 3500 m^3 of low and intermediate radioactive waste removed
16 December 1987	Biblis, Hesse, Germany	Loss of coolant accident in a reactor design system where they were not supposed to happen (1/33 million risk)
16 August 1987	Novaya Zemlya, Russia	Underground nuclear weapon test which released some radioactivity into the atmosphere
9 February 1991	Mihama, Japan	Mechanical failure and human error, leaking radioactive steam off-site in Japan's worst nuclear accident

Sources: May 1989; Oberg 1988; Sanger 1991.

Certainly two of the most important accidents occurred in the recent past: Three Mile Island, Pennsylvania, in 1979; and Chernobyl, Ukraine, in 1986. Both accidents are landmarks in the development of nuclear policy and political opposition to it. Three Mile Island is recognized not so much for the amount of radioactivity released (15–24 curies), but rather for the ensuing overhaul of the American nuclear industry after the accident. It also challenged the prevailing scientific view on the impossibility of these types of accidents, further eroding public confidence in the technology. The major safety innovations and design changes mandated after TMI effectively crippled the nuclear industry. No new reactors have been ordered or built since 1979, primarily due to the costs of the safety changes, licensing, but more importantly as a consequence of public opposition. Locally, the legacy of Three Mile Island has been governmental and industrial mistrust and the psychological stress on residents (Sills 1982; Sorenson *et al* 1987; Goldsteen and Schorr 1991).

Chernobyl was a defining event in another way. It was one of the largest and most catastrophic nuclear power plant accidents in the history of atomic power. More than 1000 miles^2 of earth are contaminated around the reactor. There was radioactive fallout throughout Europe as the radioactive plume swirled with the prevailing wind patterns. Dozens of people were killed immediately; more are languishing with thyroid cancer (especially children) and other types of malignant tumors (Gould 1991). Five years after the accident, areas of Belarus still have readings of more than 5 curies per km^2 of cesium 137 (Brooke 1991). The number of other people developing longer-term cancers not only in the former Soviet republics, but elsewhere in Europe, is unknown.

The severity of Chernobyl, the transboundary nature of its impacts, and the reluctance of the former Soviet Union to immediately acknowledge the accident and report it to the United Nations International Atomic Energy Agency (IAEA) prompted a number of institutional changes in the international governance of nuclear power plants. The first was the Nuclear Accident Notification Treaty (1986) which compels nations to provide relevant information on nuclear accidents to the IAEA in a timely manner so as to minimize the transboundary consequences of the radioactivity. The second treaty signed at the same time, the Nuclear Accident Assistance Treaty, facilitates the provision of assistance by the international community in the event of a nuclear accident. Both of these treaties were signed by more than 70 nations including all the nuclear countries.

Politically, Chernobyl created great anxiety worldwide on the safety of nuclear power plants. The concern about safety was especially heightened in eastern and central Europe where many of the reactors were built with the same design. The German government, for example, has closed five reactors in Griefswald (formerly East Germany) (World Resources Institute 1992). Inter-regional conflicts over proposed nuclear reactors in border regions are being hotly contested not only within countries but between them. The most poignant example is the construction of the Temelin reactors in Ceske Budejovice, Czechoslovakia. Austria is very much opposed to these reactors primarily because of their Russian design. The construction of the reactors (four are planned) has stimulated public protests in both countries. Austria has even offered to compensate Czechoslovakia for lost electrical power and to supply technical support for decommissioning the partially completed reactors (World Resources Institute 1992). Improving the regulation of old reactors is now part of the European Community's agenda, and they recently appropriated US $60 million for aid to Czechoslovakia, Bulgaria, and the former Soviet Union for improving the regulation of reactors (Wald 1991).

Remnant nuclear landscapes

There are certain places in the world that are still contaminated by radioactivity either as a result of an accident, or the consequence of military weapons production and testing. I call these places remnant nuclear landscapes. In his photographic essay, Goin (1990) provides the most telling images of many of these remnant nuclear landscapes (Bikini Atoll, the Nevada Test Site, the Alamagordo Bombing Range). There are many other remnant nuclear landscapes outside the US: Eniwetok, Mururoa Atoll, Pripyat, Monte Bello islands, to name just a few.

Remnant nuclear landscapes are also places where commercial reactors are located, fuel intact, but are no longer operating. They are awaiting decommissioning and just sit in a state of limbo, until such time as a final resting place for their highly radioactive cores can be found. Decommissioning removes the spent fuel, scrubs all the contaminated tubing and structural surfaces, and removes the irradiated steel and concrete from the site so it may be put to another use. As nuclear reactors age, they begin to wear out; estimates are that an average US reactor has a 30-year operating lifetime. Since many of the plants were built in the 1960s, some are reaching the end of their operating lives this decade. At least 50% of the 119 reactors currently operating will have to be decommissioned in the next 20 years (Breen 1992). In the US the following have all applied for decommissioning: Fort St. Vrain reactor in Colorado; Humboldt Bay, Vallecitos and Rancho Seco (closed by voters' referendum in 1989) in California; Fermi 1 in Michigan; Peach Bottom 1 in Pennsylvania; and Shoreham (which never came on line) in New York (USNRC 1991). Despite the ageing of the American reactors, decommissioning has not really engaged much of the public's attention (Pasqualetti 1984, 1988, 1990).

With the restructuring of the US economy into more peacetime activities, a number of the weapons complex sites will become remnant nuclear landscapes as well. The problems of ageing of the facilities themselves as well as the wastes they produce has made these places more publicly visible (Resnikoff 1990; Reicher and Scherr 1990). The Rocky Flats plant, the chemical processing plant at the Idaho National Engineering Laboratory, the Feed Materials Production Centre in Ohio, are closed and remain so, leaving behind both radioactive and non-radioactive hazardous waste (see Chapter 6). Other production reactors have been closed (notably those at Savannah River and Hanford) pending the resolution of safety concerns (Schneider 1991b).

Reducing nuclear threats

When the Atomic Energy Commission first advanced its commercial nuclear programme it had to argue the safety of the technology as the public was sceptical at the time. Popular magazines such as *National Geographic* touted the usefulness of atomic energy to fight disease, to power cars, planes, and ocean liners, to help factories and farmers, and to provide a source of industrial power (Colton 1954; Boyer 1985). Titles such as 'Man's new servant, the friendly atom' (Colton 1954), and 'You and the obedient atom' (Fisher 1958) illustrate the selling of the fledgling nuclear industry.

At the same time, the government was releasing its estimates on accident probabilities. The earliest figure (the WASH-740 report) claimed that the probability of failure was between 1/100,000 to 1/billion per year per reactor. These probabilities were subsequently revised (the range was narrowed) first in 1974 (the Rasmussen Reactor Safety Study) and a year later in another NRC study. In each case, there was no mention of the potential for a core-meltdown accident, an accident that was thought so improbable these reports did not even try to quantify its probability of occurrence. To help alleviate the damages that a failure might cause, the nuclear industry sought liability protection from the government, in the form of the Price Anderson Act, first passed in 1957. This act limited the compensation of victims of nuclear power plant accidents to US $560 million per accident. In 1987 the limit was raised to US $7 billion. The understatement of risks plus the government insurance provided the context for the nuclear industry to rapidly expand during the 1960s and 1970s.

Emergency planning

Emergency planning for nuclear accidents or nuclear war has been part and parcel of the atomic age. Civil defense programmes during the 1950s produced the public fallout shelters and the 'duck and cover' drills (when you hear the siren, crawl under your desk and cover your head) that many schoolchildren practised with the help of the cartoon figure, Bert the Turtle. How-to-protect-yourself manuals were produced for the public including designs for constructing your own fallout shelter (Science Service 1950). Civil-defense programmes went largely unchallenged until the 1980s, when the American public realized the futility in preparing for something that was not survivable, despite government opinions and manuals to the contrary (Kearney 1980). Many academics, particularly those in the hazards and disaster fields, refused to participate in these planning efforts and become increasingly critical of them (Platt 1984; Leaning and Keys 1984).

Emergency planning for nuclear power plant accidents was always a part of the licensing process for commercial reactors. In fact, many of the first reactors were intentionally located away from population centres to reduce the impacts if one were to fail. In the beginning the AEC established a 16-mile exclusion area surrounding the reactor where people could not live (Golding *et al* 1992). However, these standards were relaxed to a 2-mile zone during the 1950s based on two notions: (1) the AEC feeling that core-meltdown accidents were not credible (no probability of occurrence); and (2) containment buildings would prevent large releases from smaller accidents. By 1970 the AEC required plants to have on-site emergency plans and three years later suggested that off-site planning in conjunction with local and state officials be implemented. By 1978 a joint task force between the NRC and the USEPA established criteria for off-site planning; a 10-mile primary zone and a 50-mile secondary zone (Cutter 1984). Then Three Mile Island occurred and demonstrated that indeed a core-meltdown accident was credible. Emergency-response planning changed overnight as emergency managers and the public found out just how inadequate the planning efforts were. Planning for reactor accidents has been hotly contested with many state governments refusing to participate as a way of thwarting the licensing of the plants. If off-site plans were not developed and tested with the local governments involved, then no operating license was issued to the utility. Strategy proved successful in the case of Shoreham, New York. The lack of local and state involvement in planning (a sign of nuclear resistance) led the NRC to revise its rules and regulations to allow utility companies to prepare off-site plans with or without local cooperation. In a parallel development, model emergency plans were commissioned as part of the aftermath of the TMI accident (Golding *et al* 1992).

The impact of TMI on siting and safety issues was felt internationally as well. This was especially true in the United Kingdom where siting and safety regulations and the British nuclear reactor programme are continually challenged (Openshaw 1986; Blowers and Pepper 1987; MacGill 1987; O'Riordan, Kemp and Purdue 1988).

Arms control

The threat of nuclear proliferation has always posed problems for nuclear nations and as soon as the nuclear weapons technology was perfected, arms treaties (bilateral and multilateral) were negotiated. One of the earliest was the 1963 Partial Test Ban Treaty that banned nuclear weapons testing in the atmosphere, outer space, and under water. The US, former Soviet Union, and the UK signed the treaty as did 122 other nations. The other two original members of the nuclear club, France and China, did not sign then or now. Another important international treaty was the Non-Proliferation Treaty signed in 1968 (but effective as of 1970). This treaty attempted to prevent the spread of nuclear weapons and promoted instead the peaceful uses of nuclear energy and nuclear disarmament. More than 138 signed this treaty with some notable exceptions: Argentina, France, Brazil, China, India, Israel, Pakistan, South Africa. These nations have become, of course, the newest members of the nuclear club. France and China were already members.

Bilateral treaties between the US and the former Soviet Union characterized arms control in the 1970s as common ground was found to reduce strategic arms including antiballistic missile systems and other offensive weapons. Throughout this period and even earlier, American public opinion was somewhat ambivalent about the arms race. In the late 1940s, for example, 60% of the population felt the development of the atomic bomb was a good thing, but by the mid-1960s only 40% felt that way. By the 1970s and mid-1980s, less than 30% feel the development of the atomic bomb is good (Russett 1991). Similarly, Americans have always been quite pragmatic in their assessments of the dangers of nuclear weapons. In 1951, for example, 50% of the public felt that an atom-bomb attack was not survivable. This all changed during the 1980s as superpower tensions escalated. By 1987 more than 83% replied that they would not survive a nuclear attack.

Nuclear resistance

Waging peace

Public opposition to nuclear technology is very cyclical. Mass protests are most visible when there is a specific threat (such as a new weapon system) or when many groups (peace, environmental, women's) find a common focal point. The ineffectiveness of arms control, failures in bilateral treaty negotiations, and a rise in anti-nuclear opinion led the public to a more aggressive stance in nuclear policy decision-making. A brief synopsis of some of these popular resistance efforts is instructive, as they illustrate the effectiveness of mass movements in the political process and in determining the social acceptability of risk. In most cases the leading advocates were women or individuals outside the military and industrial power structure that controlled the technology and its uses.

Immediately after the war, some of the Manhattan Project scientists felt the need to educate the public and work toward establishing international control of atomic weapons in the search for a permanent peace. In 1947 they founded the *Bulletin of the Atomic Scientists* as a forum for public policy discussions and debates. One feature of *The Bulletin* is its doomsday clock, a symbol marking the nuclear threat and global catastrophe which stood at seven minutes to midnight. In 1957 an organization called SANE (Committee for a Sane Nuclear Policy) was formed in Washington DC, calling for nuclear disarmament and providing a non-governmental institutional base for both men and women activists (Boyer 1985). One of the earliest anti-nuclear protest movements was the Women's Strike for Peace on 1 November 1961. In response to fears that atomic weapons testing was leading to planetary destruction, more than 50,000 women in 60 American cities went on strike for one day (Swerdlow 1982). Since the women were mostly middle-class and worked in the home not outside of it, their impact was enormous. This action is credited with providing some of the impetus for the Test Ban Treaty signed in 1963 (Harris and King 1989). Other notable protest activities included the Women's Pentagon Action (1980) where feminist, environmentalist, and peace activists converged on the Pentagon to voice their opposition to the nuclear threat. This was followed by the Women's Peace Camps (Women's Encampment for a Future of Peace and Justice) which practised civil disobedience by occupying land outside of military bases first at Greenham, England (1981), and later at the Seneca depot (1983) in New York. There were continuous local demonstrations at many nuclear weapons facilities during this period as well, including the Nevada Test Site.

The failure of the Reagan administration to renew the Partial Nuclear Test Ban Treaty as well as increased defense spending for nuclear weapons again heightened public support for disarmament talks during the late 1970s. A proposal to negotiate a mutual and verifiable halt to the production, testing and deployment of nuclear weapons and their delivery systems was developed by the activist, Randall Forsberg in 1979. Known as the Nuclear Weapons Freeze, this concept garnered grass-roots political support beginning in New England in early 1982 (Cutter, Holcomb and Shatin 1986). By 12 June 1982 750,000 people converged on New York City protesting the nuclear arms race and calling for a freeze; the largest disarmament rally in the history of the US. During the autumn 1982 elections, more than 25% of the American electorate voted on the bilateral freeze initiative (Cutter, Holcomb and Shatin 1986; Cutter *et al* 1987), and by 1984 the Nuclear Weapons Freeze was a central part of the national Democratic platform and bills were making their way through Congress (Waller 1987). Though the Democrats lost the 1984 election, and the Freeze movement fizzled, it nevertheless had an enormous impact on arms control by illustrating the power of grass-roots mobilization in response to the threat of nuclear annihilation (Shah 1992). Resistance to the presence of nuclear weapons throughout the world has led to renewed social protests in a number of host nations including the formation of political parties such as Die Grünen in Germany. Kidron and Smith (1983) provide a series of maps of nuclear resistance that highlight the transnational appeal of mass protest movements at the time.

Another effective form of nuclear resistance is the declaration of nuclear-free zones. Nuclear-free zones are places (mostly political jurisdictions, but not always) where the production, transportation and storage of nuclear weapons, and the processing and disposal of nuclear materials is prohibited. International treaties have established five world regions as nuclear-free zones (Table 7.5). These include Antarctica (1959), Outer Space (1967), Latin America (1967), the international seabed (1971), and the South Pacific (1985). In addition, 26 countries have declared themselves nuclear-free either by explicitly or implicitly prohibiting nuclear weapons by law or by policy, although the policy may not be enforced (Fig. 7.8) In addition to these countries, over 4400 communities in 24 countries have declared themselves nuclear-free, including 188 in the US (Nuclear Free America 1992). Interestingly, the one community to declare itself nuclear-free in Kazakhstan, is none other than Semipalatinsk, home of research and testing of nuclear weapons.

Despite international treaties and mass protests, the dismantling of the Soviet Union has resulted in the most profound changes in the geography of nuclear hazards. In recognition of this, the *Bulletin of Atomic Scientists*, the public conscience of the nuclear threat, has reduced the time for doomsday to 17 minutes to midnight, the farthest it has been since its inception. In contrast, during the Reagan era, the clock stood at three minutes to midnight in 1984. This symbol of public opinion has been one of the most influential voices in assessing nuclear hazards (Table 7.6), mobilizing public opinion, and influencing public policy.

The long tentacle of US weapons has been withdrawn as has that of the former Soviet Union. The US is slowly withdrawing its military forces

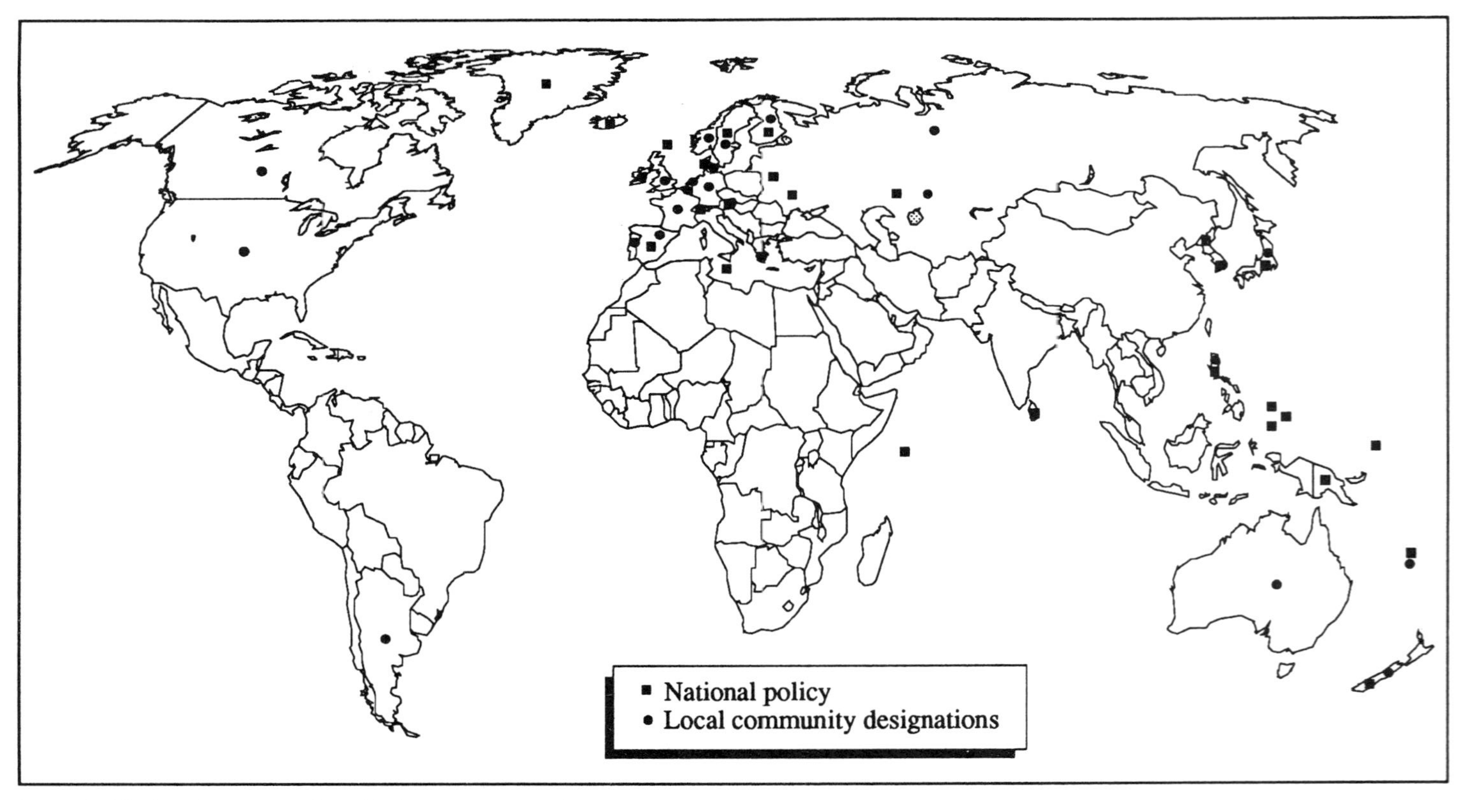

Fig. 7.8 Nuclear-free zone activity. Nuclear weapons are banned by national law, policy, or inclination in 26 countries while individual communities in 24 countries have declared themselves nuclear-free. Source: Nuclear Free America 1992.

Table 7.5 Nuclear-free zones.

Antarctica
- established by the Antarctica Treaty, 1959
- ratified by 26 nations including the US and USSR
- demilitarized the continent

Outer Space
- established by the Outer Space Treaty, 1967
- ratified by 89 nations including the US and USSR
- banned weapons of mass destruction from space

Latin America
- established by the Treaty of Tlatelolco, 1967
- ratified by 24 nations including US, USSR, China, and France; not signed by Argentina and Brazil
- banned nuclear weapons from Latin America including Cuba, Puerto Rico, and the US Virgin Islands

International Seabed
- established by the International Seabed Treaty, 1971
- ratified by 73 nations including the US and USSR

South Pacific
- established by the Treaty of Rarotonga, 1985
- signed by nine nations with two pending; protocols ratified by the USSR, China; US, France, and UK refused to sign

Source: Nuclear Free America 1989, 1992.

(and their nuclear weapons) from western Europe. Nuclear weapons are no longer based in eastern and central Europe. Four of the states of the Commonwealth of Independent States (Belarus, Ukraine, Kazakhstan, and Russia) are now nuclear nations. Yet all have agreed in principal to consolidate the weapons (at the US's insistence in return for economic aid) and place them under control of Russia, who will assume the obligations of the former USSR regarding arms-control treaties, including the responsibilities under the Non-Proliferation Treaty.

Backyards become battlefields

Another form of nuclear resistance is local opposition to the siting of nuclear facilities. Many commercial reactors and weapons complexes were established decades ago when local opposition was minimal, or were located in areas out of the public's view. The not-in-my-backyard syndrome (NIMBY) is very much a part of contemporary society, making the location of perceived noxious facilities (ranging from sanitary landfills, to prisons, to incinerators) virtually impossible in many places. As the perceived risks of the facility increase, so does local opposition to it.

The NIMBY syndrome with its exceedingly parochial outlook is most pronounced in the US and the United Kingdom, especially over the

Table 7.6 The *Bulletin of Atomic Scientists'* doomsday clock.

Date	Minutes to midnight	Reason for change
1947	7	Founding of the *Bulletin*
1949	3	USSR explodes first atomic bomb
1952	2	Successful test of US hydrogen bomb
1960	7	Growing public understanding of nuclear weapons and futility of the arms race
1963	12	Partial Test Ban Treaty signed
1968	7	France and China acquire nuclear weapons
1969	10	US Senate ratified Nuclear Non-Proliferation Treaty
1972	12	Strategic Arms Limitation Treaty (SALT I) and Anti-Ballistic Missile Treaties signed by US and USSR
1974	9	SALT II talks at an impasse; India joins nuclear weapons club
1980	7	US-USSR arms talks stalemate; Ronald Reagan elected President
1981	4	Arms race escalates as US and USSR develop more nuclear warheads
1984	3	Propaganda replaces arms-control negotiations; arms race escalates
1988	6	East-West relations improve; US and USSR sign bilateral treaty to eliminate intermediate-range nuclear forces
1990	10	Cold War ends as democratization movements take over Europe
1991	17	US and USSR sign Strategic Arms Reduction Treaty, and make cuts in tactical and strategic weapons

Source: Bulletin of the Atomic Scientists 1991.

siting of nuclear-waste repositories. Despite federal legislaton enacted in 1982 (Nuclear Waste Policy Act) and amended in 1987, the US still has been unable to site and develop a high-level waste repository, although one candidate site, Yucca Mountain, Nevada, has been selected for evaluation. Meanwhile, 1200 m^3 of high-level waste from commercial reactors is stored in congested pools at the reactors and another 400,000 m^3 of high-level waste sits at DoE/defense sites (CEQ 1992).

The issue of radioactive-waste management is a history of contested politics (Jacob 1990). There is enormous political opposition (Kraft and Clary 1991) fuelled by the public's perception of the risks which are rooted in the images of atomic bombs and mushroom-type clouds. These fears coupled with new accounts of the mishandling and contamination of the military's weapons facilities has led to enormous mistrust of the government, especially the Department of Energy (Slovic *et al* 1991a,b). The public fear, translates into local and state opposition (the NIMBY syndrome) setting the stage for confrontations. The mistrust, fear, and local opposition is no less vociferous in other nations (Blowers, Lowry and Soloman 1991; Openshaw, Carver and Fernie 1989; Kemp 1990).

In an attempt to solve the short-term problem for reactors, the US government is building a network of temporary storage sites to store the spent fuel rods until they can be permanently entombed in a high-level waste repository. Based on their siting experiences, DoE has tried a different approach for the temporary warehouse. Economic incentives, in the form of research grants are provided to communities who want to be considered as hosts for these facilities. Lured by these lucrative grants (several millions of dollars), seven communities (Yakima Indian Nation, Washington; Mescalero Apache Tribe, New Mexico; Chickasaw Indian Nation, Oklahoma; Sac and Fox Nation, Oklahoma; Fremont County, Wyoming; Grant County, North Dakota; and Prairie Island Indian Community, Minnesota) have expressed an interest (Schneider 1992). Despite local acceptance, critics argue that the government is taking advantage of rural poverty and practising racism in its quest for a place to keep the radioactive waste. There is some truth in these objections, since five of the communities are Indian nations, and all are located in rural, economically depressed counties. Thus the quest for the high-level repository continues with conflicts over federal authority and local jurisdiction, as well as scientific and public disagreements over credibility, legitimacy, and the social and economic priorities for deciding where the waste products of the peaceful atom should ultimately rest.

The situation with low-level waste and sites for its disposal is faring no better. The volume of low-level waste has steadily climbed during the 1980s from 770,000 m^3 in 1980 to more than 1.38 million m^3 in 1990, an increase of more than 80% (CEQ 1991). Constrained by NIMBY responses, Congress passed the Low Level Radioactive Waste Policy Act (LLRWPA) in 1980 requiring states to develop disposal sites by 1985 and allowing currently operating sites (Hanford, Washington, Barnwell, South Carolina, Nevada Test Site) to accept low-level waste until 1992 after which time states were on their own. When it became clear that the 1985 deadline was far too ambitious, the law was amended and the new deadline set for 1996. In 1996 radioactive waste would become the property of the state which then would have to assume liability for any damages caused by the waste. The act forced states to identify sites that would be geologically stable enough to contain the waste for 500 years. Some states entered into regional compacts and began the delicate negotiations in deciding on the host state and potential sites within it (Table 7.7).

Table 7.7 Low-level radioactive waste-disposal compacts.

Compact name	States	Host state	% Nation's low-level waste
Northwest	Washington, Oregon, Idaho, Montana, Utah Alaska, Hawaii	Washington	7
Southwestern	California, Arizona, South Dakota, North Dakota	California	9
Rocky Mountain	Wyoming, Nevada, Colorado, New Mexico	Nevada	<1
Midwest	Minnesota, Iowa, Missouri, Ohio, Wisconsin, Michigan, Indiana	Michigan	10
Central Midwest	Illinois, Kentucky	Illinois	9
Central	Nebraska, Kansas, Oklahoma, Arkansas, Louisiana	Nebraska	5
Southeast	Virginia, North Carolina, South Carolina, Florida, Georgia, Alabama, Mississippi, Tennessee	South Carolina	31
Appalachian	Pennsylvania, West Virginia, Delaware, Maryland	Pennsylvania	10
Northeast	New Jersey, Connecticut	New Jersey and Connecticut	6
Texas	Texas	Texas	1
Unaffiliated	New York	New York	6
	Massachusetts	Massachusetts	3
	Maine	Maine	<1
	Vermont	Vermont	<1
	New Hampshire, Rhode Island	–	<1

Source: USNRC 1991, p. 111.

In 1990, the Governor of New York filed a federal lawsuit claiming that the 1985 law was unconstitutional. He claimed that since 90% of the low-level waste was from nuclear reactors it should be managed by them or the federal government. It seemed fair since the federal government was the promoter of nuclear power but shunted the responsibility for disposal onto the states (Yarrow 1992). In June 1992 the US Supreme Court sided with New York and the showdown between the states and the federal government over the responsibility for managing low-level radioactive waste entered a new era.

Apocalypse now or later?

I have illustrated that the hazards of nuclear technology are still pervasive throughout the world. Despite the lessening of tensions by the former superpowers and the reductions in their nuclear arsenals, nuclear hazards are still very much a part of the global landscape. Militarism and geopolitical tensions still drive many foreign policies. While the threat of global nuclear annihilation is less now than a few years ago, the regional threats are still very real, especially as more and more nations join the nuclear club. On the commercial side, the use of nuclear power involving a new generation of reactors is in the wings. Energy trade-offs are being made between the risks of continued fossil fuel use that contributes to global environmental risks, *versus* an expanded use of nuclear power, thereby creating regional and local risks. Radioactive wastes continue to pile up, with many nations still without permanent repositories. Nations are swallowing the bitter wormwood as they struggle toward the 21st century, ultimately making tragic choices because they have no other perceived alternative.

8

Living with risks

'The formation flew backwards over a German city that was in flames. . . . When the bombers got back to their base, the steel cyclinders were taken from the racks and shipped back to the United States of America, where factories were operating night and day, dismantling the cylinders, separating the dangerous contents into minerals. Touchingly, it was mainly women who did this work. The minerals were then shipped to specialists in remote areas. It was their business to put them into the ground, to hide them cleverly, so they would never hurt anybody ever again' (Vonnegut 1969, pp. 74–75).

Despite its fictional nature, Kurt Vonnegut's *Slaughterhouse Five or the Children's Crusade* provides one interesting perspective on reducing risks and hazards—debunking the technology by getting rid of the resources that enabled the technology to succeed in the first place. Unfortunately, this management alternative is not as easily accomplished in the modern world (decommissioning a nuclear power plant, for example) as is this fictional response to weapons reductions.

Throughout this book I have tried to illustrate a number of themes that are essential in understanding environmental hazards. First and foremost, technological risks and hazards are social constructs. They are products of failures in technological systems and/or failures in political, social, and economic systems that govern the use of technology. The world will always have technocrats: people (and their policies) who try to improve the human condition by solving all of its environmental, medical, hunger, military or economic problems with bigger and better technology. Technocratic views fail to address the underlying factors that create environmental stress or hazards in the first place such as materialism, poverty, or uneven economic power, preferring to focus only on the impacts of these on society. These views also assume that technological solutions will only improve the human condition, not degrade it. It becomes society's function, then, to balance the technocratic perspective against others in determining what is an acceptable or fair risk.

The second theme I stress is that the social construction of risk and hazards leads to differing perceptions by individuals, experts, managers, and government. These risk perceptions are not only highly variable, but also volatile, and rancorous at times. The contested nature of risks and hazards ultimately leads to politicized responses. Who is to say, for example, that the expert's judgement of risks (influenced by training, environmental philosophy, and race) is any better than an activist's (equally influenced by experience, wealth, and proximity to the source of threat)? Both are important to policy decisions as they clearly embody different definitions and characterizations of risk.

This leads to the third theme, namely that risks and hazards often become so politicized that it constrains their management. The identification, assessment, and measurement of risks and hazards, and the selection of mitigation techniques is often played out in the court of public opinion. Seemingly technical issues (this obviously depends on your perspective) become part of the public discourse. More often than not, issues of scientific uncertainty and equity dominate the debates in societal decisions regarding risk-acceptability and public policy choices.

Lastly, there are uneven burdens of risk and hazards (people and places that produce them as well as people and places affected by them) resulting in uneven responses. There are clear social, intergenerational, and regional inequities in hazard burdens. These inequities are also seen in the responses to risk and hazards, as they restrict the range of mitigation options. I described some of these in the case studies of chemical hazards, hazardous-waste flows, and nuclear technology.

It is the interaction between nature, society, and technology at a variety of spatial scales that creates the mosaic of risks or 'hazardscape' that affect places and the people who live there. The hazardscape can be the landscape of many hazards (pollution, nuclear power plants, chemical factories) within a region, or comparisons of one type of hazard between regions such as the examples in Chapters 5, 6 and 7. It is the difference between the hazardscape of New Jersey and the hazardscape of worldwide hazardous-waste flows. As you have seen, scale adds yet another complexity to the perception, assessment, acceptability, and ultimate management of risks and hazards. As you can see geography is not a trivial matter. It plays a major role in our understanding of the impacts or risks and hazards and our collective responses to them. What geographical theories are important in enhancing our understanding or risks and hazards?

From case studies to theory

One of the liveliest debates in the hazards community at present is over the development of hazards theory. First, is there an overarching theory of environmental hazards that encompasses hazard events regardless of their origins in natural, technological or social phenomena? Second, what is the relationship between hazards theory and more general questions of nature–society interactions? Despite its 50-year history, natural hazards research has provided a wealth of case study data on the mapping, perception and responses to natural events, but has been less successful in developing broader theories of nature–society interactions. Technological hazards research, on the other hand, is quite sophisticated in its theoretical development drawing from many fields such as organizational behaviour, psychology, locational analysis and communications. However, there are fewer case studies so we really have very little 'hard data' on which to judge our theories at the present time. We know even less about hazards that could arguably fall between these two such as global warming.

The natural hazards paradigm

The natural hazards paradigm developed by White, Burton and Kates originated in the human–ecological tradition within geography (Barrows 1923). The relationship between people and their environment is viewed as a series of adjustments in both the human use and natural events systems. Hazards are connected to the geophysical processes that initiate them, e.g. an earthquake or a hurricane. It is the interaction of this extreme event with the human conditions in particular places that produces the hazard and influences responses to it. Risk is synonymous with the distribution of these extreme events or natural features that give rise to them (Emel and Peet 1989). Much of the early hazards work mapped the location of these extreme events (floods, earthquakes, hurricanes) to delineate risks. But these early case studies also mapped the human occupancy of these hazard-prone areas (flood plains, seismic zones, coastal zones). In putting these together researchers could examine the adjustments people made in response to the hazards at these specific locales. Since the methodology was often the same, case studies were compared to see if similar patterns of risk, response and adjustment arose.

During the 1970s, radical critiques of this 'traditional natural hazards' view were plentiful. Arguing that the perception of and adjustments to hazards are constrained by cultural, economic, political and social forces, these researchers (Waddell 1977; Torry 1979) contend that the geography of social relations governed the spatial extent and occurrence interval of natural disasters. This critique of the causal mechanism of natural hazards—one that went beyond the geophysical characteristics of the event—became known as the political–economic (or political ecology) view of hazards. This mode of explanation is best exemplified by the work of Watts (1983), Wisner, O'Keefe and Westlake (1977), and Hewitt (1983) who illustrate that hazards were intensified and people made more vulnerable to them by political and societal constraints on human responses.

The evolution and maturation of the radical critique has definitely improved our understanding of hazards. In her review essay, Marston (1983) describes how traditional natural hazards, managerialism and political economy all offer different explanations for the increasing losses from extreme natural events at the global level, as well as proposing different solutions. She notes the insularity of these approaches and suggests that political considerations may in fact be the mechanism to integrate these perspectives into a new theory of hazards.

Hazards in context

Hazards in context also begins with the human–ecological approach but expands the natural hazards paradigm to include the social and political contexts within which the hazard takes place, a direct result of the radical critiques. Hazards in context pursues its goal through a combination of research methodologies utilizing empirical and behavioural analyses. The two most recent examples of this line of approach are Palm (1990), and Mitchell, Devine and Jagger (1989).

Palm provides a framework that focuses attention on the various levels of social, political and economic structures that constrain or enable our understanding of environmental hazards. She calls them micro, meso, and macro elements and they simply refer to individuals or households (micro), managers (meso) and the state (macro). The hazards-in-context approach seeks to reinterpret the nature–society interaction as a dialogue between the physical setting, political–economic context, and the role and influence of individuals or agents in affecting change. This complex web of relationships is spatially linked helping to further understand the dynamic status of nature–society interactions as one shifts from level to level. How changes in the political–economic systems encourage changes in the use of the physical environment or govern responses to it, and how changes in the population itself result in different environmental goals, uses and responses are key to this hazards-in-context approach. Geographic scale is a central tenet.

To illustrate this model, Palm examines the housing market's response to earthquake risk. She first identifies the physical risk, the distribution of fault zones. Constraints on the purchasing of earthquake insurance at the household level (micro), and the non-acceptance of mandated insurance at the regional level (macro) illustrate the lack of responsiveness to the hazards. As she concludes, '. . . the environment and its hazards are downplayed and ignored as much as possible. The efforts of individuals to increase the awareness and response of Californians to earthquake hazards have been largely ineffective, primarily because they run counter to the prevailing cultural values and interests of powerful actors within the political economy' (1990, p.144).

These power brokers include real-estate agents who intervene at the meso scale. Palm provides empirical proof on the importance of these agents in influencing the buyer's perception of risks which in turn affects their decision to purchase earthquake insurance. The complexity of responses to earthquake risk and the intervening factors constraining or enabling the responses at all levels (micro, meso, macro) are important contributions of Palm's case study.

In a later article, the household decision to purchase earthquake insurance, is used to clarify the micro- and macro-level influences on decision-making (Palm and Hodgson 1992). Despite state-mandated disclosure of earthquake risk and the availability of earthquake insurance, less than half of California homeowners have it. Palm and Hodgson found that insurance purchase was not spatially related to geophysical risk nor to the socioeconomic characteristics of home buyers. Rather, the key determinant in the insurance-purchasing decision was the perception of risk by the homeowners. Interestingly, there was little relationship between geophysical risk and perceived risk. Those that felt more vulnerable (a greater likelihood of serious damage from a major earthquake) were more likely to purchase earthquake insurance.

Along similar lines, Mitchell, Devine and Jagger (1989) provide a case study of the devastating wind damage resulting from coastal storms that struck the United Kingdom in 1988. Again arguing from a human–ecological vantage point, they claim their hazards-in-context model is more dynamic than previous natural hazards work; incorporating more interactive components between physical processes, population vulnerability, adjustments and losses; the main elements in most natural hazards systems. They go on to suggest that these hazard components do not occur in spatial, temporal, social, economic, or political isolation. Hazards are embedded in larger societal structures and these must be examined as well in order to fully comprehend the nature of the hazard and responses to it.

For example, the 1988 coastal windstorm caused extensive local damage throughout the southern portion of the United Kingdom, yet the event did not precipitate any significant changes in hazard management or policies. It was placed very low on the priority list of other social problems requiring public policy attention. Windstorms (and the extensive loss of trees) were simply not salient enough to public and private agendas to prompt any changes in hazard reduction and mitigation policies, despite the enormity of impacts at the local level. In both examples, the hazards in context conceptualization is an improvement on both the natural hazards paradigm and radical critiques of it. Unfortunately there are few, if any studies on technological hazards using this approach.

Social theory and hazards

As applied to hazards, social theory is the critical evaluation of the possibilities for harm from hazards and the avoidance of harm. It is an outgrowth of the radical critiques, yet moves beyond the Marxist dialectic by claiming that hazards are socially constructed, a further refinement of the hazards-in-context view. The focus is on those social structures that initiate individual responses to hazards, how those social structures are interpreted both by individuals and collections of individuals, and how these are all linked in time and space. Nature and, by analogy, natural hazards are also viewed as social constructions (Fitzsimmons 1989). The acceptance of social theory as a perspective for understanding hazards entails ideological struggles over the meaning of nature and advocacy for political change in order to control the ways in which technology and capitalism destroy the environment. As Fitzsimmons (1989) argues, 'If we do not engage in these struggles, we abandon them to those who use Nature to justify not only the domination of nature by humans, but also the domination of humankind itself' (p.117)

Kirby (1990) suggests hazards theory needs to refocus to incorporate the social construction of hazards, their historical antecedents, and the institutions that govern the management of hazards—a merger of hazards in context and social theory. Kirby's contribution to social theory and hazards is the introduction of past contexts into the discourse. Clearly, when dealing with technological hazards, this historical view is essential in understanding the range and extent of the hazard and responses to it. The role of social theory and hazards is also illustrated by the case studies presented by Johnson and Covello (1989) who examine the social and cultural construction of risk.

In the disaster-research field in sociology, social theory is also gaining popularity. In his innovative work on the Bhopal tragedy, Bogard (1989) proposes two hypotheses incorporating social theory and hazards. First, he suggests there is an increase in the hazardousness of everyday environments. Second, there is a heightened vulnerability of the poor and disadvantaged classes to hazards that lead to increased damage from actual events. Both people and places are becoming more vulnerable as a result of increasingly complex technology, and a false perception that technology has infallible safeguards. There is also an increasingly uneven distribution of hazards that affect social groups differentially, resulting in uneven vulnerability as well.

In applying this framework to the analysis of the Bhopal chemical release, release, Bogard reconstructs events leading up to the 1984 accident. He suggests that the production of this chemical hazard was related to modernization efforts which superimposed a complex production system on a society that was unable to adequately cope with it. Component failures, lax enforcement, errors in judgement, and lack of public knowledge, were all constrained by economic and political systems. The extreme conflict between the dependency on the technology (need for foreign investment, need for the pesticide product) and constraints in detecting the potential hazards helped to create the accident. Under these circumstances, according to Bogard, the accident was bound to happen. The appearance of safety (limited by the possibilities for detection of chemical releases), permitted the settlement of people close to the plant, thereby increasing the hazard and the population's vulnerability to it.

In earlier work Bogard (1988) utilizes social theory to illustrate that unacknowledged conditions and unanticipated consequences of rational action are partially responsible for the increase in damages and casualties from hazard events rather than a simple increase in the magnitude of the hazard itself. In other words, the social context is as much a contributor to the increased impacts of hazards on people and places as is the magnitude of the event itself. Clearly in the example of Bhopal this is quite true.

Social amplification of risks

Quite a different intellectual tradition is found in the risk community (see Chapter 1). In their pioneering social amplification of risks model, Kasperson *et al* (1988) suggest that risks interact with psychological, social, cultural, and institutional processes in ways that either amplify or attenuate public response. The important link here is between the technical assessments and the broader-based socially defined perspectives on risk and behaviour. The technical assessment of risk, for example, only examines the probability of events and the magnitudes of consequences, thereby ignoring questions of when, where, and who ultimately bears the burden of risk. The basic question underlying the social of amplification of risk, is why does a minor risk (as defined by technical experts) produce such massive public reactions?

Kasperson *et al* suggest that the social construction of risk leads to individual and group perceptions of the risk as well as the perceived effects of these risks on the community and society as a whole. In other words, despite the 'expert judgements', society still feels a particular product, activity, or technology is problematic, necessitating some form of response. The social amplification of risk thus becomes a corrective mechanism whereby technical risk (we could easily substitute geophysical risk) is brought more in line with its social definition. A two-stage process ensues: (1) the transfer of information about the risk to the public; and (2) feedback loops where individual and societal-response mechanisms are adjusted and sometimes altered. All of this occurs within broader political, social, and economic frameworks.

A risk event occurs such as an airline crash, a new study on the dangers of 'Alar', or a chemical factory explosion. Risks are then amplified by large volumes of information (media attention), disputes over factual information (scientific experts challenging one another's data), dramatization of the issue (television footage and other visuals), information channels (electronic and print media, personal networks), and finally through value-laden terminology and images such as mushroom clouds, toxic dumps, and skull and crossbones.

The interpretation of these risk events by society is governed by a number of factors: values (which risks are more important and the desired course of action); social group relations and their political prowess; the

ıe (significance) of the risk event or technology in question; and
gative imagery) associated with the risk (e.g. Love Canal, Three
d, Chernobyl, Bhopal). These factors interact to create a ripple
impact of the risk begins with the initial victims and diffuses
ɔ society at large. Just as a pebble thrown into a lake creates
does the amplification of risks.

There have been some critiques of this social amplification model (Rappaport 1988; Rayner 1988; Rip 1988) as well as others who have sought to improve upon it (Machlis and Rosa 1990). Inasmuch as this approach is relatively new, more empirical verification of the model and its elements is bound to be forthcoming, especially by members of the Clark University team that proposed it.

Hazards of place and vulnerability

Social theory, hazards in context, and the social amplification of risks are relatively new approaches to conceptualizing risk and hazards. They have many more commonalities than differences. One such common element is vulnerability. Vulnerability is a central theme in hazards research, yet there is very little consensus on its meaning or exactly how to assess it. The term is so ubiquitous, yet everyone has a slightly different interpretation of its meaning. Furthermore, questions of geographic scale and social referent (individual, household, community, society) add to the confusion. Are we talking about vulnerable people, places, or societies and at what scale: local, national, regional, global? Hazards researchers continually use the vulnerability concept, yet precise operational definitions are sparse. There are even fewer empirical analyses of vulnerability to specific hazards.

There are as many different interpretations of vulnerability as there are hazards researchers. Timmerman (1981), for example, defined vulnerability as 'the degree to which a system may react adversely to the occurrence of a hazardous event' (p.17). He goes on to suggest that the degree and quality of the adverse reaction is conditioned by a system's resilience, a measure of its capacity to absorb and recover from the event. Susman, O'Keefe and Wisner (1984) offer another interpretation of the concept, based in political-economic ideology. To them, vulnerability is the degree to which different classes of society are differentially at risk. A third perspective examines vulnerability as the potential for loss. Reductions in vulnerability can be accomplished through changes in societal institutions such as physical infrastructure (buildings), economic systems, and sociopolitical systems (Cuny 1983). Vulnerability as synonymous with loss potential is found in both the hazards in context (Mitchell, Devine and Jagger 1989) and social theory (Bogard 1989) approaches.

In its broadest terms, vulnerability can be defined as the likelihood that an individual or group will be exposed to and adversely affected by a hazard. Measures of vulnerability help illustrate how people are placed in hazardous situations and how they respond to them. Vulnerability studies vary widely in style and content, ranging from largely descriptive to theoretical; involve acute to chronic events and risk; and examine natural to technological hazards. The concepts of risk, preparedness, mitigation, and the social geography of affected populations are commonly found interwoven in vulnerability studies.

Most of the vulnerability research to date focuses on natural hazards or global change and either examines vulnerable places based on biophysical or environmental conditions; or vulnerable people (or societies) governed by political economic conditions. Examples of these are readily found in the natural hazards literature (Hewitt 1983; Liverman 1990a; Wilhite and Easterling 1987). There are far fewer studies in the technological area and of those, most only examine the technological conditions or failures imparting risk to places (Gabor and Griffith 1980; Pijawka and Radwan 1985; Liverman 1986; Cutter and Solecki 1989; Cutter and Tiefenbacher 1991).

Hewitt and Burton (1971) were early pioneers in the development of the vulnerability concept by examining the range of hazards affecting a particular place and responses to it. This 'hazardousness of place' model urged researchers to become knowledgeable about particular places and the totality of hazards that could or did affect it; in effect an assessment of place vulnerability. However, the regional ecology of hazards never really took hold. Whether this was due to the narrowness of researchers (who often studied only one type of hazard such as earthquakes, floods, hurricanes), or the inability of researchers to focus intensely on one place is unclear. Unfortunately, regional hazards studies never found a home in the research community, unlike regional specializations within geography.

Borrowing from the regional ecology idea of Hewitt and Burton (1971), Cutter and Solecki (1989) proposed a hazards-of-place model. The hazards of place combine with social and political structures to create vulnerable places and vulnerable people within those places. Using airborne chemical releases as an example, we suggested the degree of chemical hazard (or place vulnerability) is a function of the level of risk and the amount of pre-impact planning and mitigation in response to such risks. Furthermore, risks and mitigation occur within broader social contexts as well. At the county level, social vulnerability is a function of demographic indicators (gender, race, income, education, age, etc.), mobility/transportation indicators and the level of risk perception/communication. It is the interaction of the hazards of place (risk and mitigation) with the social profile of communities that provides an overall indicator of the vulnerability of places to airborne toxic releases. Ideally, vulnerability studies are best measured at the local level (county or community).

The Cutter and Solecki model goes beyond the original mapping technique of Hewitt and Burton, providing a more comprehensive assessment of risk, mitigation, and exposure at the local level. As yet, however, it has only been tested with one type of hazard and only at the state and metropolitan levels. Other applications of the hazards-of-place model include a recent study of pesticide-drift hazards at the county scale (Tiefenbacher 1992). The hazards-of-place model is easily expanded to include all the hazards that affect a particular locale; the beginnings of a local or regional hazardscape.

The frequent target of vulnerability studies (people *versus* places) often ignores the underlying process that creates vulnerability. As O'Riordan (1986) suggests, vulnerability must be viewed as an interactive and dynamic process that links environmental risks and society. It becomes more of a comparative measure rather than an absolute one. The integration of biophysical (or technological) conditions of risk with the political and economic conditions within which the risk occurs is the focus of much of the more innovative work in the field (Blaikie and Brookfield 1987; Bogard 1989; Liverman 1990b).

These integrative approaches to vulnerability provide the most promising new directions in hazards research. The coupling of the biophysical or technological sources of risk in time and space helps delineate the hazardscape. However, we also need to examine who is most affected by the risk and why, including how society views risks and the social equity—or more likely inequity—in risk distribution and responses.

Here again, geography matters. A more detailed examination of these hazards at a smaller spatial scale is warranted as it provides a more accurate representation of the geographic distribution of risks and the factors underlying it. We can also go into more detail concerning the affected populations and those who bear the greatest risk burdens, such as the poor, minorities, and women. At its best, vulnerability incorporates elements from all of the theoretical approaches described above in order to understand the dynamic interplay between physical, technological, and social systems.

Revisionist views

Since we all view the world through different prisms, our own view of the environment, science, and technology will affect our individual and collective responses to the risks and hazards that each poses. Feminist perspectives offer new insights into living with risks but have been largely overlooked and undervalued.

Feminism confronts science and technology

The interest in gender and science got its start in the contemporary women's movement when feminists began questioning the uses and abuses to which science has been put by men, especially in the form of patriarchal knowledge. Some of the earliest liberation struggles, ones that continue to this day, have been on regaining knowledge and control over women's bodies (health and reproduction), from the paternal scientists and practitioners. The return to more holistic medicine and natural childbirth is a rejection of the sanitized and technological approach to health care and the risks of modern medicine. Feminist critiques have helped us to understand that science is just one way of looking at the world. Science is a socially produced body of knowledge, just like other forms of knowledge that also have inherent biases. They have also helped us to recognize that a diversity of approaches in problem solving is important (Harding 1991; Bleir 1986).

Feminism has also helped to influence our view of the environment and the technology to manipulate it (see Chapter 1). Men in decision-making positions have traditionally undervalued the nurturing dimension that forms the central tenet of ecocentrism and feel more comfortable with the rational, scientific view embodied in technocentrism (Chapter 2). Technology is the creation of science and engineering, fields where women have traditionally had limited access and professional acceptance. The growing sophistication of hazards management especially during the last decade has meant the increasing separation of the manager from the public or constituency.

Policy-makers become increasingly dependent on scientific advisors, usually representing a narrow range of specialties, most of whom were and still are men. Women are continually marginalized, taking 'secondary roles' as citizen activists. Despite the work of such notable scientists as Rachel Carson's *Silent Spring*, women's views about risks and hazards are ridiculed and discounted. The separation of 'informed judgements' by experts from those of so-called 'uninformed public' fuel policy debates on the acceptability of environmental risks as we have seen (Chapter 3). It is in fact, the creation and maintenance of the aura of 'expertise' that is part of the problem.

Liberating risk and hazards research

Feminist perspectives offer a number of new insights into how we live with risks. These include: (1) how risks and hazards are identified and assessed; (2) broadening the public discourse on uneven or disproportional burdens; (3) politicizing action in influencing acceptability judgements; and (4) outright resistance to new technology and existing risks and hazards.

We have already seen that women often have very different perceptions of risks and hazards, especially those risks involving children's health (the 'Alar' example in Chapter 2). The controversy over biotechnology and genetic engineering is another good example. In 1987 the bovine growth hormone (BGH) was to be one of the first of biotechnology's large money-makers, by boosting milk production in dairy animals (Sun 1989). Genetic-engineering critics quickly mobilized grass-roots opposition arguing three main points. First, the use of BGH would drive out the family farm, an idyllic remnant of the American culture. Second, health questions about the safety of milk produced by BGH-treated cows were unresolved. Based on its own research, the Food and Drug Administration had already given approval for its use. Finally, the general hormonal nature of the product created a scare (regardless of contrary evidence) in consumers who prefer their milk 'natural'. Children and pregnant women are the largest sector of the milk-consuming public and the spectre of a major consumer boycott quickly eliminated the market debut of BGH. The potential American consumer boycott in addition to European Community bans on the use of BGH left this new biotech discovery in the laboratories.

Women are often less willing to accept certain risks and are continually leading the grass-roots efforts to reduce local risks be it from hazardous waste, nuclear technology, or biotechnology. Women often provide the first wave of opposition to new technology, either as vocal opponents or concerned consumers. The case of 'Ice Minus' provides a good example (Krimsky and Plough 1988; Piller 1991). 'Ice Minus' is a genetically engineered bacterium designed to reduce frost damage to plants and marketed under the name 'Frostban'. Proposals to test 'Ice Minus' in an open field in Tulelake, California, in 1982 incurred enormous local opposition as people were concerned about the release of a genetically altered bacterium into an uncontrolled environment. Concerned about the risks, two women spearheaded these resistance efforts and with the assistance of the biotech critic, Jeremy Rifkin, slapped a lawsuit against the University of California, halting the experiment. In 1987 'Frostban' (the product name for the 'Ice Minus' bacterium) was finally tested at Tulelake where it prevented the formation of frost as low as 23°F.

The rancorous public debate during the five-year period from application for testing to the actual test illustrate the social amplification of risks as well as the problems of risk communication, and regulatory review and management. Public concern over other R-DNA strains released outside controlled laboratory settings has led to calls for clear and enforceable regulations (Pimentel 1989).

Perhaps the most significant contribution of feminism has been its empowerment of women to take the political actions illustrated above. By personalizing environmental risks and hazards and showing the burden of environmental abuse on people, women have been and continue to seek a greater role in the public discourse and acceptability decisions. Grass-roots mobilizations stem from very personal politics. Petra Kelly's concern for the lunacy of weapons of mass destruction on her native German soil that eventually led to anti-nuclear and pro-environment political action, or Gwyn Kirk's stance against nuclear warheads in Greenham

Common, or Lois Gibbs' campaign to show the state of New York that there was a serious health problem in Love Canal, all attest to a feminist praxis: you can't separate the personal from the political.

The same pattern is true in environmental risks in developing countries. Women provide the food and fetch water for the families and collect wood for fuel. They see the results of environmental degradation first hand, e.g. they must walk farther to collect wood, food becomes scarcer, and children suffer as a result of contaminated water. Environmental resistance such as the Chipko movement in India (women 'tree-huggers' who halted the deforestation of their land), or the Greenbelt movement (reforestation) in Kenya started by the National Council of Women (Wangari Maathai) illustrate the effectiveness and potential for grass-roots mobilization in response to environmental risks and hazards.

Risks and hazards are embedded in our political, economic, and social institutions. We should learn from our past mistakes and failures and embrace the diversity of contested and politicized views of risk and hazards. This will ultimately lead to improved management strategies and reductions in risk and hazards. While we can learn to adjust and live with risks and hazards, it doesn't always mean that we have to.

References

Abelson, P.H. 1992. Remediation of hazardous waste sites, *Science* 255 (5047), 901.
Aftalion, F. 1991. *A History of the International Chemical Industry*. Philadelphia: University of Pennsylvania Press.
Ainsworth, S. 1991. Tighter curbs sought on transport of chemicals, *Chemical and Engineering News* 69(34), 5–6.
Ainsworth, S. and Lepkowski, W. 1991. Metham-sodium spill shows tankcar safety flaws, *Chemical and Engineering News* 69(30), 7–8.
Ajzen, I. and Fishbein, M. 1980. *Understanding Attitudes and Predicting Social Behavior*. Englewood Cliffs, NJ: Prentice-Hall.
Arcury, T.A., Scollay, S.J. and Johnson, T.P. 1987. Sex differences in environmental concern and knowledge: the case of acid rain, *Sex Roles* 16, 463–472.
Arkin, W.M. and Fieldhouse, R.W. 1985. *Nuclear Battlefields: Global Links in the Arms Race*. Cambridge, MA: Ballinger.
Asch, P. 1990. Food safety regulation: is the Delaney clause the problem or symptom? *Policy Sciences* 23, 97–110.
Auty, R.M. 1984. The product life-cycle and the location of global petrochemical industry after the second oil shock, *Economic Geography* 60(4), 325–338.
Barrows, H.H. 1923. Geography as human ecology, *Annals of the Association of American Geographers* 12, 1–14.
Bartimus, T. and McCartney, S. 1991. *Trinity's children: Living Along America's Nuclear Highway*. New York: Harcourt Brace Jovanovich.
Bastide, S., Moatti, J.P. and Fagnani, F. 1989. Risk perception and social acceptability of technologies: the French case, *Risk Analysis* 9, 215–223.
Baumann, D.D. and Sims, J.H. 1972. The tornado threat: coping styles of the north and south, *Science* 176, 1386–1392.
Bazelon, D.L. 1979. Risk and responsibility, *Science* 205, 277–280.
Berry, B.J.L. 1977. *The Social Burdens of Environmental Pollution*. Cambridge, MA: Ballinger.
Blaikie, P.M. and Brookfield, H.C. 1987. *Land Degradation and Society*. London: Methuen.
Bleir, R. (ed.) 1986. *Feminist Approaches to Science*. New York: Pergamon.
Blowers, A., Lowry, D. and Solomon, B.D. 1991. *The International Politics of Nuclear Waste*. New York: St. Martin's Press.
Blowers, A. and Pepper, D. (eds) 1987. *Nuclear Power in Crisis*. London: Croom Helm.
Bogard, W.C. 1988. Bringing social theory to hazards research: conditions

and consequences of the mitigation of environmental hazards, *Sociological Perspectives* 31(2), 147–168.

Bogard, W.C. 1989. *The Bhopal Tragedy: Language, Logic, and Politics in the Production of Hazard*. Boulder: Westview Press.

Boyer, P. 1985. *by the Bombs Early Light: American Thought and Culture at the Dawn of the Atomic Age*. New York: Pantheon.

Boulding, E. 1984. Focus on: the gender gap, *Journal of Peace Research* 21, 1–3.

Breen, B. 1992. Dismantling nuclear reactors, *Garbage* 4(2) March/ April, 40–47.

Brinkman, R, Jasanoff, S. and Ilgen, T. 1985. *Controlling Chemicals: The Politics of Regulation in Europe and the United States*. Ithaca: Cornell University Press.

Britton, N.R. and Oliver, J. 1991. *Natural and Technological Hazards: Implications for the Insurance Industry*. Armidale, Australia: University of New England. Proceedings of a Seminar sponsored by Sterling Offices (Australia) Ltd.

Broad, W.J. 1991. A Soviet company offers nuclear blasts for sale to anyone with the cash, *The New York Times* 7 November, A18.

—— 1992. In Russia, secret labs struggle to survive, *The New York Times* 14 January, C1.

Brodeur, P. 1989. *Currents of Death*. New York: Simon and Schuster.

Brooke, J. 1991. Chernobyl said to affect health of thousands in a Soviet region, *The New York Times*, 3 November, A1.

Brown, M.H. 1987. *The Toxic Cloud*. New York: Harper and Row.

—— 1980a. Drums of death, *Audubon* 82(4), 120–133.

—— 1980b. *Laying Waste: Love Canal and the Poisoning of America*. New York: Pantheon.

Bryant, B. and Mohai P. (eds) 1992. *Race and the Incidence of Environmental Hazards: A Time for Discourse*. Boulder: Westview Press.

Bullard, R.D. 1990. *Dumping in Dixie: Race, Class, and Environmental Quality*. Boulder: Westview Press.

Bulletin of the Atomic Scientists 1991. December 47(10).

Burton, I. 1962. Types of Agricultural Occupance of Flood Plains in the US. Research Paper No. 75. Chicago: University of Chicago Department of Geography.

Burton, I., Kates, R.W. and White, G.F. 1978. *The Environment as Hazard*. New York: Oxford University Press.

Cashmann, J.R. 1988. *Hazardous Materials Emergencies: Response and Control (Second Edition)*. Lancaster, PA: Technomic.

Center for Investigative Reporting and Bill Moyers 1990. *Global Dumping Ground: The International Traffic in Hazardous Waste*. Washington: Seven Locks Press.

Chapman, K. 1991. *The International Petrochemical Industry*. Oxford: Blackwell.

—— 1992. Continuity and contingency in the spatial evolution of industries: the case of petrochemicals. *Transactions Institute of British Geographers* 17(1), 47–64.

Chemical Week 1983. Restructuring: how the chemical industry is building its

future. *Chemical Week* 26 October, 26–62.
—— 1988. Hazardous waste management: putting solutions into place, *Chemical Week* 143(8) 24 August, 26–58.
—— 1991a. Forecast 1991, *Chemical Week* 2 January/9 January 1991, 14–30.
—— 1991b. Hazardous waste, shrinking options–tough choices, *Chemical Week* 149(1) 21 August, 40–57.
—— 1992. CW300, 4 March, 24.
Chepesiuk, R. 1991. From ash to cash: the international trade in toxic waste, *E Magazine* 11(4) July/August, 30–37.
Chess, C. and Hance, B.J. 1989. Opening doors: making risk communication agency reality, *Environment* 31(5), 11–15, 38–39.
Church, A.M. and Norton, R.D. 1981. Issues in emergency preparedness for radiological transportation accidents, *Natural Resources Journal* 21, 757–771.
Chynoweth, E. 1991. EC emergency response review, *Chemical Week* 148(23), 12.
Clarke, L. 1988a. Politics and bias in risk assessment, *Social Science Journal* 25(2), 155–165.
—— 1988b. Explaining choices among technological risks, *Social Problems* 35(1), 22–35.
Colton, C.E. 1990a. Environmental developments in the East St. Louis region 1890–1970, *Environmental History Review* 14(1–20), 93–114.
—— 1990b. Historical hazards: the geography of relict industrial wastes, *The Professional Geographer* 42(2), 143–156.
Colton, F.B. 1954. Man's new servant, the friendly atom, *National Geographic* January, 71–90.
Comfort, L.K. (ed.) 1988. *Managing Disaster: Strategies and Policy Perspectives*. Durham: Duke University Press.
Commission for Social Justice 1987. *Toxic Waste and Race in the United States*. New York: United Church of Christ.
Committee on Earth Sciences, US Geological Survey 1990. *Our Changing Planet: the FY 1991 US Global Change Research Program*. Reston, VA: US Geological Survey.
—— 1991. *Our Changing Planet: the FY 1992 US Global Change Research Program*. Reston, VA: US Geological Survey.
—— 1992. *Our Changing Planet: the FY 1993 US Global Change Research Program*. Reston, VA: US Geological Survey.
Commoner, B. 1989. The hazards of risk assessment, *Columbia Journal of Environmental Law* 14(2), 365–378.
Congressional Quarterly Service. Congressional Quarterly Almanac 1979–1990.
Congressional Quarterly Service 1991. Congressional Weekly Reports, 1991.
Council on Environmental Quality 1991. *Environmental Quality: 21st Annual Report*. Washington DC: Government Printing Office.
—— 1992. *Environmental Quality: 22nd Annual Report*. Washington DC: Government Printing Office.
Covello, V.T. 1983. The perception of technological risks: a literature review, *Technological Forecasting and Social Change* 23, 285–297.
Covello, V.T. and Allen, F. 1988. Seven cardinal rules of risk communication. Washington DC: USEPA Office of Policy.

Covello, V.T. and Mumpower, J. 1985. Risk analysis and risk management: an historical perspective, *Risk Analysis* 5(2), 103–120.

Covello, V.T., Sandman, P.M. and Slovic, P. 1988. *Risk Communication, Risk Statistics and Risk Comparisons: A Manual for Plant Managers.* Washington DC: Chemical Manufacturers Association.

Crouch, E.A.C. and Wilson, R. 1982. *Risk/Benefit Analysis.* Cambridge, MA: Ballinger.

Cuny, F.C. 1983. *Disasters and Development.* New York: Oxford University Press.

Cutter S.C. 1978. *Community Attitudes Toward Pollution.* Chicago: University of Chicago Department of Geography Research Monograph No. 188.

Cutter, S.C. 1981. Community concern for pollution: social and environmental influences, *Environment and Behavior* 13(1), 105–124.

Cutter, S.L. 1984. Residential proximity and cognition of risk at Three Mile Island: implications for evacuation planning. In Pasqualetti, M. and Pijawka, K.D. (eds), *Nuclear Power: Assessing and Managing Hazardous Technology.* Boulder: Westview Press, p. 247–258.

—— 1984. Emergency preparedness and planning for nuclear power plant accidents, *Applied Geography* 4, 235–245.

—— 1987. Airborne toxic releases: are communities prepared? *Environment* 29(6), 12–17, 28–31.

—— 1991. Fleeing from harm: international trends in evacuations from chemical accidents, *International Journal of Mass Emergencies and Disasters* 9(2), 267–285.

—— 1992. Toxic monuments and the making of technological hazards, in Janelle, D. et al (eds), *Geographical Snapshots of North America.* New York: Guilford Press, pp. 117–121.

—— and Barnes, K. 1982. Evacuation behavior and Three Mile Island, *Disasters* 6(2), 116–124.

—— and Solecki, W.D. 1989. The national pattern of airborne toxic releases, *The Professional Geographer* 41(2), 149–161.

—— and Tiefenbacher, J. 1991. Chemical hazards in urban America, *Urban Geography* 12(5), 417–30.

—— Tiefenbacher, J. and Solecki, W.D. 1992. En-gendered fears: femininity and technological risk perception, *Industrial Crisis Quarterly* 6, 5–22.

——, Holcomb, H.B. and Shatin, D. 1986. Spatial patterns of support for a nuclear weapons freeze, *Professional Geographer* 38(1), 42–52.

——, Holcomb, H.B., Shatin, D., Shelley, F.M. and Murauskas, G.T. 1987. From grass roots to partisan politics: nuclear freeze referenda in New Jersey and South Dakota, *Political Geography Quarterly* 6(4), 287–300.

Daggett, C.J., Hazen, R.E. and Shaw, J.A. 1989. Advancing environmental protection through risk assessment, *Columbia Journal of Environmental Law* 14(2), 315–328.

Davenport, J.A. 1977. A survey of vapor cloud incidents, *Chemical and Engineering Process* 73(9), 54–63.

Davis, L.N. 1984. *The Corporate Alchemists.* New York: William Morrow and Company.

DePol, D.R. and Cheremisinoff, P. 1984. *Emergency Response to*

Hazardous Materials Incidents. Lancaster, PA: Technomic Publishing Co.
Dickens, C. 1978. *The Uncommercial Traveller and Reprinted Pieces Etc.* Oxford: Oxford University Press.
Dietz, T. and Rycroft, R.W. 1987. *The Risk Professionals*. New York: Russell Sage Foundation.
Dietz, T., Frey, R.S. and Rosa, E. 1983. Risk, technology, and society, in Dunlap, R.E. and Michelson, W. (eds), *Handbook of Environmental Sociology*. Westport, CT: Greenwood Press.
Douglas, M. and Wildavsky, A. 1982. *Risk and Culture: An Essay on the Selection of Technological and Environmental Dangers*. Berkeley: University of California Press.
Downs, A. 1972. Up and down with ecology—the 'issue-attention cycle', *Public Interest* 28, 38–50.
Drabek, T.E. and Hoetmer, G.J (eds) 1991. *Emergency Management: Principles and Practice for Local Government*. Washington DC: International City Management Association.
Dunlap, R.E. 1991a. Trends in public opinion toward environmental issues: 1965–1990, *Society and Natural Resources* 4, 285–312.
—— 1991b. Public opinion in the 1980s: clear consensus, ambiguous commitment, *Environment* 33(8), 10–15, 32–37.
—— and Scarce, R. 1991. The polls-poll trends: environmental problems and protection, *Public Opinion Quarterly* 55, 651–672.
Edelstein, M. 1988. *Contaminated Communities: The Social and Psychological Impacts of Residential Toxic Exposure*. Boulder: Westview Press.
Ellul, J. 1964. *The Technological Society*. New York: Vintage Books.
Emel, J. and Peet, R. 1989. Resource management and natural hazards, in Peet, R. and Thrift, N. (ed.), *New Models in Geography, Volume One*. London: Unwin Hyman, pp. 49–76.
Englander, T., Farago, K., Slovic, P. and Fischhoff, B. 1986. A comparative analysis of risk perception in Hungary and the United States, *Social Behavior* 1, 55–66.
European Chemical News 1988. Italy recalls toxic waste ships, 12 September 1988.
Fawcett, H.H. and Woods, W.S. (eds) 1982. *Safety and Accident Prevention in Chemical Operations*. New York: John Wiley and Sons.
Finkel, A.M. 1989. Is risk assessment really too conservative? Revising the revisionists, *Columbia Journal of Environmental Law* 14(2), 427–468.
Fischer, G.W., Morgan, M.G., Fischhoff, B., Nair, I. and Lave, L.B. 1991. What risks are people concerned about? *Risk Analysis* 11(2), 303–324.
Fischhoff, B. 1985. Managing risk perceptions, *Issues in Science and Technology* II (1), 83–96.
Fischhoff, B., Lichtenstein, S., Slovic, P.L., Derby, S.L., Keeney, R.L. 1981. *Acceptable Risk*. New York: Cambridge University Press.
Fischhoff, B., Slovic, P. and Lichtenstein, S. 1979. Weighing the risks, *Environment* 21(4), 17–20, 32–38.
Fischhoff, B., Slovic, P., Lichtenstein, S., Read, S. and Combs, B. 1978. How safe is safe enough? A psychometric study of attitudes towards technological risks and benefits, *Policy Sciences* 9, 127–152.
Fisher, A.C., Jr. 1958. You and the obedient atom, *National Geographic*

September, 303–352.
FitzSimmons, M. 1989. The matter of nature, *Antipode* 21(2), 107–120.
Ford, D. 1982. *The Cult of the Atom*. New York: Simon and Schuster.
Foster, H.D. 1980. *Disaster Planning: the Preservation of Life and Property*. New York: Springer-Verlag.
Freudenburg, W.R., 1988. Perceived risk, real risk: social science and the art of probabilistic risk assessment. *Science* 242, 44–49.
Friedman, T.L., 1992. Reducing the Russian arms threat, *The New York Times* 17 June.
Fuller, J.G. 1977. *The Poison that Fell from the Sky*. New York: Berkeley Medallion.
Gabor, T. and Griffith, T.K. 1980. The assessment of community vulnerability to acute hazardous materials incidents, *Journal of Hazardous Materials* 0, 323–333.
Gardner, G.T. and Gould, L.C. 1989. Public perceptions of the risks and benefits of technology, *Risk Analysis* 9, 225–242.
Gibbs, L.M. 1982. *Love Canal: My Story as told to Murray Levine*. New York: Grove Press.
Goin, P. 1990. *Nuclear Landscapes*. Baltimore: Johns Hopkins University Press.
Golden, T. 1992. At least 180 dead as blasts rock a Mexican city, *The New York Times*, 23 April, Al.
Golding, D., Kasperson, J.X., Kasperson, R.E., Goble, R., Seley, J.E., Thompson, G. and Wolf, C.P. 1992. *Managing Nuclear Accidents: A Model Emergency Response Plan for Power Plants and Communities*. Boulder: Westview Press.
Goldman, B., Hulme, J.A., Johnson, C. 1986. *Hazardous Waste Management: Reducing the Risk*. Washington DC: Island Press.
Goldsteen, R.L. and Schorr, J.K. 1991. *Demanding Democracy After Three Mile Island*. Gainesville: University of Florida Press.
Goldstein, B.D. and Greenberg, M. 1991. Environmental applications and interventions in public health, in Holland W.W., Detels, R., and Knox, G. (eds), *Oxford Textbook of Public Health Volume 3, Applications in Public Health*. Oxford: Oxford University Press, pp. 17–28.
Gould, J.M. and Goldman, B.A. 1991. *Deadly Deceit: Low-level Radiation High-level Cover-up*. New York: Four Walls Eight Windows Press.
Gould, L.C., Gardner, G.T., DeLuca, D.R., Tiemann, A.R., Doob, L.W. and Stolwijk, J.A.J. 1988. *Perceptions of Technological Risks and Benefits*. New York: Russell Sage Foundation.
Gould, P. 1991. *Fire in the Rain: The Democratic Consequences of Chernobyl*. Baltimore: The Johns Hopkins University Press.
Greenberg, M.R. 1986. Health effects of environmental chemicals, *Journal of Planning Literature* 1(1), 1–13.
——, Sachsman, D.B., Sandman, P.M. and Salomone, K.L. 1989. Network evening news coverage of environmental risk, *Risk Analysis* 9(1), 119–126.
Greenhouse, S. 1988. Toxic waste boomerang: ciao Italy!, *The New York Times* 3 September.
Greenpeace 1990. *The International Trade in Wastes: A Greenpeace Inventory (5th Edition)*. Washington DC: Greenpeace USA.
Haber, L.F. 1971. *The Chemical Industry 1900–1930: International Growth and Technological Change*. Oxford: Clarendon Press.

Hadden, S.G. 1989. *A Citizen's Right to Know: Risk Communication and Public Policy*. Boulder: Westview Press.
Hall, B. and Kerr, M.L. 1991. *1991–92 Green Index: A State-by-State Guide to the Nation's Environmental Health*. Washington DC: Island Press.
Hammitt, J.K. 1990. Risk perceptions and food choice: an exploratory analysis of organic versus conventional-produce buyers, *Risk Analysis* 10(3), 367–374.
Hanson, D. 1991. Safety rules expected for railcars carrying hazardous materials, *Chemical and Engineering News* 69(4), 16–17.
Hansson, S.O. 1989. Dimensions of risk, *Risk Analysis* 9, 107–112.
Hardin, C.M. and Eiser, J.R. 1984. Characterizing the perceived risk and benefits of some health issues, *Risk Analysis* 4(2), 131–141.
Hardin, G. 1968. The tragedy of the commons. *Science* 162, 1243–1248.
Harding, S. 1991. *Whose Science? Whose Knowledge? Thinking from Women's Lives*. Ithaca: Cornell University Press.
Harris, A. and King, Y. (eds) 1989. *Rocking the Ship of State: Toward a Feminist Peace Politics*. Boulder: Westview Press.
Harris, R.C., Hohenemser, C. and Kates, R.W. 1978. Our hazardous environment, *Environment* 20(7), 6–15, 38–41.
Hayes, D. 1990. Harnessing market forces to protect the earth, *Issues in Science and Technology* VII (2) Winter, 46–51.
Headrick, D.R. 1990. Technological change, in Turner, B.L. *et al* (eds), *The Earth as Transformed by Human Action*. Cambridge: Cambridge University Press, 55–67.
Heaney, P. 1992. People and plumbism: cultural perceptions of lead exposure. New Brunswick, NJ: Department of Geography, Rutgers University Discussion Paper.
Heller, K. 1991. Buffer zones head for the border, *Chemical Week* 149(4), 22.
—— 1992. Ads, advocacy, outreach activists, *Chemical Week* 150(23), 19–24.
Hester, G.L. 1992. Electric and magnetic fields: managing an uncertain risk, *Environment* 34(1), 6–11, 25–32.
Hewitt, K. and Burton, I. 1971. The Hazardousness of a Place: A Regional Ecology of Damaging Events. Toronto: University of Toronto Department of Geography Research Publication No. 6.
Hewitt, K (ed.) 1983. *Interpretations of Calamity*. Winchester, MA: Allen and Unwin.
Hohenemser, C., Kates, R.W. and Slovic, P. 1983. The nature of technological hazard, *Science* 220, 378–384.
Hohenemser, C., Kates, R.W., Slovic, P. 1985. A causal taxonomy, in Kates R.W., Hohenemser, C. and Kasperson, R.E., *Perilous Progress: Managing the Hazards of Technology*. Boulder: Westview Press, pp. 67–90.
Huber, P.W. 1986. The Bhopalization of American tort law, in National Academy of Engineering, *Hazards: Technology and Fairness*, Washington: National Academy Press, pp. 89–110.
Industrial Economics, Inc. 1989. *Acute Hazardous Events Data Base*. Final Report for the Office of Policy Analysis, US Environmental Protection Agency (EPA 68-W8-0038). Cambridge, MA: Industrial Economics, Inc.
Intergovernmental Panel on Climate Change (IPCC) 1990. *Climate Change: The IPCC Scientific Assessment* (ed. by Houghton, J.T., Jenkins, G.J., and Ephraums, J.J.). Cambridge: Cambridge University Press.

—— 1991. *Climate Change: The IPCC Response Strategies*. Washington DC: Island Press.
Jacob, G. 1990. *Site Unseen: The Politics of Siting a Nuclear Waste Repository*. Pittsburgh: University of Pittsburgh Press.
—— and Kirby, A. 1990. On the road to ruin: the transport of military cargoes, in Ehrlich, A.H. and Birks, J.W. (eds), *Hidden Dangers: Environmental Consequences of Preparing for War*. San Francisco: Sierra Club Books, pp. 71–95.
Janis, I.L. and Mann, L. 1977. *Decision Making: A Psychological Analysis of Conflict, Choice, and Commitment*. New York: The Free Press.
Jasanoff, S. 1987. EPA's regulation of Daminozide: unscrambling the messages of risk. *Science, Technology and Human Values* 12(3/4), 116–124.
—— 1990. *The Fifth Branch: Science Advisers as Policymakers*. Cambridge: Harvard University Press.
Johnson, B.B. 1985. Congress as hazard maker, in Kates, R.W., Hohenemser, C. and Kasperson, R.E., *Perilous Progress: Managing the Hazards of Technology*. Boulder: Westview Press, pp. 455–475.
Johnson, B.B. and Covello, V.T. (eds) 1989. *The Social and Cultural Construction of Risk: Essays on Risk Selection and Perception*. Dordrecht: D. Reidel Publishing Company.
Johnson, E.J. and Tversky, A. 1983. Affect, generalization and the perception of risk, *Journal of Personality and Social Psychology* 45, 20–31.
—— 1984. Representations of perceptions of risk, *Journal of Experimental Psychology: General* 113(1), 55–70.
Jordan, W.A. 1989. Exciting '50s, *Chemical Week* 2 August, 44–56.
Kahneman, D., Slovic, P. and Tversky, A. (eds) 1982. *Judgement Under Uncertainty: Heuristics and Biases*. New York: Cambridge University Press.
Kasperson, J.X. and Kasperson, R.E. 1987. Priorities in profile: managing risks in developing countries, *Risk Abstracts* 4(3), 113–118.
Kasperson, R.E., Kates, R.W. and Hohenemser, C. 1985. Hazard management, in Kates, R.W., Hohenemser, C. and Kasperson, J.X. (eds), *Perilous Progress*, pp. 43–66.
Kasperson, R.E. Renn, O., Slovic, P. *et al* 1988. The social amplification of risk: a conceptual framework, *Risk Analysis* 8(2), 177–187.
Kasperson, R.E. and Stallen, P.J.M. (eds) 1991. *Communicating Risks to the Public: International Perspectives*. Dordrecht: Kluwer Academic.
Kates, R.W. 1971. Natural hazard in human ecological perspective: hypotheses and models, *Economic Geography* 47, 438–451.
Kates, R.W. and Kasperson, J.X. 1983. Comparative risk analysis of technological hazards (a review), *Proc. Natl. Acad. Sci* 80, 7027–7038.
Kates, R.W. 1978. *Risk Assessment of Environmental Hazard*. New York: John Wiley and Sons, SCOPE 8.
Kates, R.W. 1986. Managing technological hazards: success, strain, and surprise, in National Academy of Engineering (ed.), *Hazards: Technology and Fairness*. Washington: National Academy Press, pp. 206–220.
Kates, R.W., Hohenemser, C and Kasperson, J.X. 1985. *Perilous Progress: Managing the Hazards of Technology*. Boulder: Westview Press.
Kearney, C. 1980. *Nuclear War Survival Skills*. Naperville: Caroline House.
Kemezis, P. 1991. OCAW tags OSHA settlement a 'sweetheart deal', *Chemical*

Week 149(16), 21.
—— 1991a. Waste not wanted, but Alabama must take it, *Chemical Week* 148(23), 19 June, 9.
—— 1991b. Utah approves another hazwaste burner—the second in the west, *Chemical Week* 149(16) 13 November, 12.
—— 1992. Among the states: free trade, *chemical week* (151(7) 19 August, 50–52
Kemp, R. 1990. Why not in my backyard? A radical interpretation of public opposition to the deep disposal of radioactive waste in the United Kingdom, *Environment and Planning A* 22, 1239–1258.
Keown, C.F. 1989. Risk perceptions of Hong Kongese *vs.* Americans. *Risk Analysis* 9, 401–405.
Kempton, W. 1991. Lay perspectives on global climate change, *Global Environmental Change Human and Policy Dimensions* 1(3), 183–208.
Kidron, M. and Smith, D. 1983. *The War Atlas: Armed Conflict-Armed Peace.* New York: Simon and Schuster.
Kirby, A. (ed.) 1990. *Nothing to Fear: Risks and Hazards in American Society.* Tucson: University of Arizona Press.
Kishchuk, N. 1987. Causes and correlates of risk perception: a comment, *Risk Abstracts* 4(1), 1–3.
Kleindorfer, P.R. and Kunreuther, H.C. 1987. *Insuring and Managing Hazardous Risks: From Seveso to Bhopal and Beyond.* Berlin: Springer-Verlag.
Kleinhesselink, R.R. and Rosa, E.A. 1991. Cognitive representation of risk perceptions: a comparison of Japan and the United States, *Journal of Cross-Cultural Psychology* 22(1), 11–28.
Kraft, M.E. and Clary, B.B. 1991. Citizen participation and the NIMBY syndrome: public response to radioactive waste disposal, *The Western Political Quarterly* 44(2), 299–328.
Krimsky, S. and Plough, A. 1988. *Environmental Hazards: Communicating Risks as a Social Process.* Dover, MA: Auburn House.
Kunreuther, H., Desvouges, W.H. and Slovic, P. 1988. Nevada's predicament: public perceptions of risk from the proposed nuclear waste repository, *Environment* 30(8), 16–20, 30–33.
Lagadec, P. 1982. *Major Technological Risk: An Assessment of Industrial Disasters.* Oxford: Pergamon.
Lathrop, J.W. (ed.) 1981. *Planning for Rare Events: Nuclear Accident Preparedness and Management.* Oxford: Pergamon.
Lave, L. 1982. Methods of risk assessment, in Lave, L. (ed.), *Quantitative Risk Assessment in Regulation.* Washington DC: The Brookings Institution, pp. 23–54.
—— 1987. Health and safety risk analysis: information for better decisions, *Science* 236, 291–295.
—— 1989. Risk assessment and regulatory priorities, *Columbia Journal of Environmental Law* 14(2), 307–314.
Leaning J. and Keyes L. (eds) 1984. *The Counterfeit Ark.* Cambridge, MA: Ballinger.
Lepkowski, W. 1991. OSHA fines Union Carbide $2.8 million, *Chemical and Engineering News* 69(38), 6.
Levin, I.P., Snyder, M.A. and Chapman, D.P. 1988. The interaction of

experimental and situational factors and gender in a simulated risky decision-making task, *The Journal of Psychology* 122, 173–181.

Levine, G. 1982. *Love Canal: Science, Politics and People*. Lexington, MA: Lexington Books.

Lewis, H.W. 1990. *Technological Risk*. New York: Norton.

Lichtenstein, S., Slovic, P., Fischhoff, B., Layman, M. and Combs, B. 1978. Judged frequency of lethal events, *Journal of Experimental Psychology: Human Learning and Memory* 4(6), 551–581.

Lindell, M.K. and Earle, T. 1983. How close is close enough: public perception of the risk of industrial facilities, *Risk Analysis* 3(4), 245–253.

Lindell, M.K. and Perry, R.W. 1990. Effects of the Chernobyl accident on public perceptions of nuclear power plant accidents, *Risk Analysis* 10(3), 393–399.

Liverman, D. 1986. The vulnerability of urban areas to technological risks, *Cities* May, 142–147.

—— 1990a. Vulnerability to global environmental change, in Kasperson, R.E., Dow, K. Golding, D. and Kasperson, J.X. (eds), *Understanding Global Environmental Change: The Contributions of Risk Analysis and Management*. Worcester, MA: The Earth Transformed Program, Clark University, pp. 27–44.

—— 1990b. Drought in Mexico: climate, agriculture, technology and land tenure in Sonora and Puebla, *Annals of the Association of American Geographers* 80(1), 49–72.

Lotstein, E.L. 1990. Recreational behavior in a barrier island park. Unpublished Ph.D dissertation, Department of Geography, Rutgers University, New Brunswick, NJ.

Lowrance, W.W. 1976. *Of Acceptable Risk: Science and Determination of Safety*. Los Altos, CA: William Kaufmann.

Lowry, G.G. and Lowry, R.C. 1988. *Lowry's Handbook of Right-to-Know and Emergency Planning*. Chelsea, MI: Lewis Publishers.

Lyman, F. 1990. *The Greenhouse Trap*. Boston: Beacon Press.

MacGill, S.M. 1987. *The Politics of Anxiety: Sellafield's Cancer Link Controversy*. London: Pion.

McCay, B.J. and Acheson, J.M. (eds) 1987. *The Question of the Commons*. Tucson: The University of Arizona Press.

McDaniels, T.L. 1988. Comparing expressed and revealed preferences for risk reduction: different hazards and question frames, *Risk Analysis* 8(4), 593–604.

McStay, J.R. and Dunlap, R.E. 1983. Male-female differences in concern for environmental quality, *International Journal of Women's Studies* 6, 291–307.

Macey, S.M. and Brown, M.A. 1983. Residential energy conservation through repetitive household behavior, *Environment and Behavior* 15, 123–141.

Machlis, G.E. and Rosa, E.A. 1990. Desired risk: broadening the social amplification of risk framework, *Risk Analysis* 10(1), 161–168.

Maderthaner, R., Guttman, G., Swaton, E. and Otway, H. 1978. Effects of distance upon risk perception, *Journal of Applied Psychology* 63(3), 380–382.

Markusen, A.R. 1985. *Profit Cycles, Oligopoly, and Regional Development*. Cambridge: MIT Press.

Marshall, E. 1991. A is for apple, Alar and . . .alarmist? *Science* 254 (4 October), 20–22.

Marshall, V.C. 1987. *Major Chemical Hazards*. New York: Halstead.
Marston, S.A. 1983. Natural hazards research: towards a political economy perspective, *Political Geography Quarterly* 2(4), 339–348.
Marx, J. 1990. Animal carcinogen testing challenged, *Science* 250, 743–745.
May. J. 1989. *The Greenpeace Book of the Nuclear Age: The Hidden History, The Human Cost*. New York: Pantheon.
May, P.J. and Williams, W. 1986. *Disaster Policy Implementation: Managing Programs Under Shared Governance*. New York: Plenum.
Mazmanian, D. and Morell, D. 1992. *Beyond Superfailure: America's Toxic Policy for the 1990s*. Boulder: Westview Press.
Merchant, C. 1981. Earthcare, *Environment* 23, 6–13, 38–40.
Milbrath, L. 1984. *Environmental Vanguard for a New Society*. Albany: State University of New York Press.
Miller, T.A.W. and Keller, E.B. 1991. What the public thinks, *EPA Journal* 17(2), 40–42.
Mitchell, J. 1992. Chemical industry of the former USSR, *Chemical and Engineering News* 70(15), 46–66.
Mitchell, J.K., Devine, N. and Jagger, K. 1989. A contextual model of natural hazard, *Geographical Review* 79(4), 391–409.
Mofson, P. 1992. Personal communication, US Department of State.
Mohai, P. 1990. Black environmentalism, *Social Science Quarterly* 71(4), 744–765.
Morehouse, W. and Subramaniam, M.A. 1986. *The Bhopal Tragedy: What Really Happened*. New York: Council of International and Public Affairs.
Morgan, M.G., Slovic, P., Nair, I. *et al* 1988. Powerline frequency electric and magnetic fields: a pilot study of risk perception, *Risk Analysis* 5(2), 139–150.
Morgenstern, R. and Sessions, S. 1988. EPA's unfinished business, *Environment* 30(6), 14–17, 34–39.
Morris, G., 1991a. Pesticide spill may stir DOT rules, *Chemical Week* 149(1), 9.
—— 1991b. Louisiana enacts environment scorecard, *Chemical Week* 149(4), 9.
Mounfield, P.R. 1991. *World Nuclear Power*. London: Routledge.
Nash, J.R. 1976. *Darkest Hours*. Chicago: Nelson-Hall.
National Academy of Engineering 1986. *Hazards: Technology and Fairness*. Washington: National Academy Press.
National Academy of Engineering, Institute of Medicine 1991. *Policy Implications of Greenhouse Warming*. Washington DC: National Academy Press.
National Research Council 1987. *Regulating Pesticides in Food. The Delaney Paradox*. Washington DC: National Academy Press.
—— 1989. *Improving Risk Communication*. Washington DC: National Academy Press.
National Research Council 1983. *Risk Assessment in the Federal Government: Managing the Process*. Washington DC: National Academy Press.
New Jersey Department of Environmental Protection and Energy 1991. *Site Remediation Program Site Status Report*. Trenton: NJDEPE.
Nimmo, D. and Combs, J. 1985. *Nightly Horrors: Crisis Coverage in Television Network News*. Knoxville: University of Tennessee Press.
Nriagu, V.O. 1983. *Lead and Lead Poisoning in Antiquity*. New York: John Wiley & Sons.

Nuclear Free America 1989. Nuclear free zones in the world and nuclear free zones in the US, *The New Abolitionist* VII (3,4), 9–12.
—— 1992. Personal communication with Chuck Johnson.
Nuclear News 1991. Map sheets of US and World commercial nuclear reactors.
Oberg, J.E. 1988. *Uncovering Soviet Disasters: Exploring the Limits of Glasnost*. New York: Random House.
Openshaw, S. 1986. *Nuclear Power: Siting and Safety*. London: Routledge & Kegan Paul.
Openshaw, S., Carver, S. and Fernie, J. 1989. *Britain's Nuclear Waste: Safety and Siting*. London: Belhaven Press.
Organization for Economic Cooperation and Development 1987. *Environmental Data Compendium*. Paris: OECD.
—— 1989. *Environmental Data Compendium*. Paris, OECD.
—— 1991a. *The State of the Environment*. Paris: OECD.
—— 1991b. *Environmental Data*. Paris: OECD.
—— 1991c. *Environmental labelling in OECD countries*. Paris: OECD.
O'Riordan, T. 1981. *Environmentalism*. London: Pion.
—— 1986. Coping with environmental hazards, in Kates, R.W. and Burton, I. (eds), *Geography, Resources, and Environment. Volume II. Themes from the Work of Gilbert F. White*. Chicago: The University of Chicago Press, pp. 272–309.
—— 1988. The politics of environmental regulation in Great Britain, *Environment* 30(8), 5–9, 39–44.
O'Riordan, T., Kemp, R. and Purdue, M. 1988. *Sizewell, B: An Anatomy of the Inquiry*. London: Macmillan.
Otway, H. and Thomas, K. 1982. Reflections on risk perception and policy, *Risk Analysis* 2, 69–82.
Otway, H. and von Winterfeldt, D. 1982. Beyond acceptable risk: on the social acceptability of technologies, *Policy Sciences* 14, 247–256.
Palm, R.I. 1990. *Natural Hazards: An Integrative Framework for Research and Planning*. Baltimore: Johns Hopkins University Press.
—— and Hodgson, M. 1992. Earthquake insurance: mandated disclosure and homeowner response in California, *Annals of the Association of American Geographers* 82(2), 207–222.
Pasqualetti, M. 1984. The decommissioning dilemma, *Sierra* (September/October), 64–70.
—— 1988. Decommissioning at ground level, *Land Use Policy* 5(1), 45–61.
—— (ed.) 1990. *Nuclear Decommissioning and Society: Public Links to a New Technology*. London: Routledge.
Perrow, C. 1984. *Normal Accidents: Living with High-Risk Technologies*. New York: Basic Books.
Perry, R.W., Lindell, M.K. and Greene, M.R. 1981. *Evacuation Planning in Emergency Management*. Lexington, MA: Lexington Books.
Pijawka, K.D. and Radwan, A.E. 1985. The transportation of hazardous materials: risk assessment and hazard management, in *Dangerous Properties of Industrial Materials Report* September/October, 2–11.
Piller, C. 1991. *The Fail-Safe Society*. New York: Basic Books.
Pimentel, D., Hunter, M.S., La Gro, J.A. *et al* 1989. Benefits and risks of genetic

engineering in agriculture, *Bioscience* 39(9), 606–614.
Pine, C. 1991. *Liability Issues*, in Drabek, T.E. and Hoetmer, G.J. (eds), *Emergency Management: Principles and Practice for Local Government*. Washington DC: International City Management Association, pp. 289–307.
Platt, R.H. 1984. The planner and nuclear crisis relocation, *Journal of the American Planning Association* 50, 259–260.
Popper, F.J. 1983. LP/HC and LULUs: The political uses of risk analysis in land-use planning, *Risk Analysis* 3(4), 255–263.
Portney, P.R. (ed.) 1990. *Public Policies for Environmental Protection*. Washington DC: Resources for the Future.
Potter, W.C. 1991. Russia's nuclear entrepreneurs. *The New York Times*, November 7, A29.
Quarantelli, E.L. and Dynes, R.R. 1977. Response to social crisis and disaster, *Annual Review of Sociology* 3, 23–49.
Quarantelli, E.L. 1987. Disaster studies: an analysis of social historical factors affecting the development of research in the area, *International Journal of Mass Emergencies and Disasters* 5(3), 285–310.
Quirk, J. and Terasawa, K. 1981. Nuclear regulation: an historical perspective, *Natural Resources Journal* 21 (October), 833–855.
Rappaport, R. 1988. Toward postmodern risk analysis, *Risk Analysis* 8(2), 189–191.
Rayner, S. 1988. Muddling through metaphors to maturity: a commentary on Kasperson *et al*, the social amplification of risk. *Risk Analysis* 8(2), 201–204.
Rayner, S. and Cantor, R. 1987. How fair is safe enough? The cultural approach to societal technology choice, *Risk Analysis* 7(1), 3–9.
Reich, M.R. 1991. *Toxic Politics: Responding to Chemical Disasters*. Ithaca: Cornell University Press.
Reicher, D.W. and Scherr, S.J. 1990. The bomb factories: out of compliance and out of control, in Ehrlich, A.H. and Birks, J.W. (eds), *Hidden Dangers: Environmental Consequences of Preparing for War*. San Francisco: Sierra Club Books, pp. 35–49.
Renn, O. and Swaton, E. 1984. Psychological and sociological approaches to study risk perception, *Environment International* 10, 557–575.
Resnikoff, M. 1990. The generation time-bombs: radioactive and chemical defense wastes, in Ehrlich, A.H. and Birks, J.W. (eds), *Hidden Dangers: Environmental Consequences of Preparing for War*. San Francisco: Sierra Club Books, pp. 18–34.
Rip, A. 1988. Should social amplification of risk be counteracted? *Risk Analysis* 8(2) 193–197.
Roberts, L. 1990. Risk assessors taken to task. *Science* 247, 1173.
Roder, W. 1961. Attitude and knowledge on the Topeka flood plain, in White, G.F. (ed.), *Papers on Flood Problems*. Research Paper No. 70. Chicago: University of Chicago Department of Geography.
Rosen, J.D. 1990. Much ado about Alar, *Issues in Science and Technology* VII (Fall), 85–90.
Rotman, D. 1990. Hazardous waste: tightening rules—and options—up the ante, *Chemical Week* 147(7) 22 August, 34–44.
—— 1992. Viewpoint. *Chemical Week* 151(7) 19 August, 4.

Rowe, W.D. 1977. *An Anatomy of Risk*. New York: Wiley
Rummel-Bulska, I. and Osafo, S. (eds) 1991. *Selected Multi-lateral Treaties in the Field of the Environment, Volume 2*. Cambridge: Grotius Publications Ltd for UNEP.
Russell, C.S. 1990. Monitoring and enforcement, in Portney, P.R. (ed), *Public Policies for Environmental Protection*. Washington DC: Resources for the Future. pp. 243–274.
Russell, M. and Gruber, M. 1987. Risk assessment in environmental policy-making, *Science* 236, 286–290.
Russett, B. 1991. Doves, hawks, and US public opinion, *Political Science Quarterly* 105(4), 515–538.
Saarinen, T.F. 1966. Perception of the Drought Hazard on the Great Plains. Research Paper No. 106. Chicago: University of Chicago Department of Ccography.
Sagers, M.J. and Shabad, T. 1990. *The Chemical Industry in the USSR: An Economic Geography*. Boulder: Westview Press.
Salzhauer, A.L. 1991. Obstacles and opportunities for a consumer ecolabel, *Environment* (9), 10–15, 33–37.
Sandman, P., Sachsman, D., Greenberg, M. and Gochfeld, M. 1987. *Environmental Risk and the Press*. New Brunswick: Transaction Publishers.
Sanger, D.E. 1991. Japan now tells of radiation release, *The New York Times* February 12, A3.
Schahn, J. and Holzer, E. 1990. Studies of individual environmental concern: the role of knowledge, gender, and background variables, *Environment and Behavior* 22(6), 767–786.
Schneider, K. 1991a. Minorities join to fight polluting of neighborhoods. *The New York Times* 25 October, A20.
—— 1991b. US plans big cuts in its production of nuclear arms, *The New York Times* 17 December, A1.
—— 1992. Grants open doors for nuclear waste, *The New York Times* 9 January, A14.
Schneider, S.H. 1989a. The greenhouse effect: science and policy, *Science* 243, 771–781.
——1989b. *Global Warming: Are We Entering the Greenhouse Century?*. San Francisco: The Sierra Club.
Schuck, P.H. 1987. *Agent Orange on Trial: Mass Toxic Disasters in the Courts*. Cambridge: Belknap Press.
Schweitzer, G.E. 1991. *Borrowed Earth, Borrowed Time: Healing America's Chemical Wounds*. New York: Plenum.
Schwing, R. and Albers, W. Jr. (eds) 1980. *Societal Risk Assessment: How Safe is Safe Enough?* New York: Plenum.
Science Service 1950. *Atomic Bombing: How to Protect Yourself*. New York: Wm. H. Wise & Co.
Scott, R.M. 1989. *Chemical Hazards in the Workplace*. Chelsea, Michigan: Lewis Publishers.
Seigel, L., Cohen, G. and Goldman, B. 1991. *The US Military's Toxic Legacy*. Boston: The National Toxic Campaign Fund.
Shah, S. (ed.) 1992. *Between Fear and Hope: A Decade of Peace Activism*.

Baltimore: Fortkamp Publishers.
Shapiro, M. 1990. Toxic substances policy, in Portney, P.R. (ed.), *Public Policies for Environmental Protection*. Washington DC: Resources for the Future. pp. 195–241.
Shippee, G., Burroughs, J. and Wakefield, S. 1980. Dissonance theory revisited, *Environment and Behavior* 12(1), 33–51.
Short, J.F. 1984. The social fabric at risk: toward the social transformation of risk analysis, *American Sociological Review* 49, 711–725.
Shrader-Frechette, K.S. 1991. *Risk and Rationality*. Berkeley: University of California Press.
Shrivastava, P. 1987. *Bhopal: Anatomy of a Crisis*. Cambridge: Ballinger.
Shulman, S. 1992. *The Threat at Home: Confronting the Toxic Legacy of the US Military*. Boston: Beacon Press.
Sills, D.L., Wolf, C.P. and Shelanski, V.B. (eds) 1982. *Accident at Three Mile Island: The Human Dimension*. Boulder: Westview.
Silverman, J.M. and Kumka, D.S. 1987. Gender differences in attitudes toward nuclear war and disarmament, *Sex Roles* 16, 189–203.
Slovic, P. 1987. Perception of risk, *Science* 236 (4799), 280–285.
—— Fischhoff, B. and Lichenstein, S. 1980. Facts and fears: understanding perceived risk, in Schwing, R. and Albers, W. (eds), *Societal Risk Assessment: How Safe is Safe Enough?* New York: Plenum, pp. 181–214.
—— 1985. Characterizing perceived risk, in Kates, R.W., Hohenemser, C. and Kasperson, J.X. (eds), *Perilous Progress: Managing the Hazards of Technology*. Boulder: Westview Press, 91–125.
Slovic, P., Flynn, J.H. and Layman, M. 1991a. Perceived risk, trust, and the politics of nuclear waste, *Science* 254, 1603–1607.
Slovic, P., Layman, M. and Flynn, J.H. 1991b. Risk perception, trust and nuclear waste: Lessons from Yucca Mountain, *Environment* 33(3), 6–11, 28–36.
Smets, H. 1987. Compensation for exceptional environmental damage, in Kleindorfer, P.R. and Kunreuther, H.C. (eds), *Insuring and Managing Hazardous Risks: from Seveso to Bhopal and Beyond*. Berlin: Springer-Verlag, 79–144.
Solecki, W.D. 1990. Acute chemical disasters and rural United States hazardscapes. Unpublished dissertation, Department of Geography, Rutgers University.
Sorenson, J.H. 1987. Evacuations due to off-site releases from chemical accidents: experience from 1980 to 1984, *Journal of Hazardous Materials* 14, 247–257.
—— and Rogers, G.O. 1988. Local preparedness for chemical accidents: a survey of US communities, *Industrial Crises Quarterly* 2, 89–108.
Sorenson, J.H., Rogers, G.O. and Glevenger, W.F. 1988. Review of public alert systems for emergencies at fixed chemical facilities. Oak Ridge, TN: Oak Ridge National Laboratory, ORNL/TM-10825.
Sorenson, J.H., Vogt, B. and Mileti, D. 1987. Evacuation: an assessment of planning and research. Oak Ridge, TN: Oak Ridge National Laboratory, ORNL-6376.
Sorenson, J.H., Soderstrom, J., Copenhaver, E., Carnes, S. and Bolin, R. 1987. *Impacts of Hazardous Technology: The Psycho-Social Effects of Restarting TMI-1*. Albany: State University of New York Press.
Spitz, P.H. 1988. *Petrochemicals: the Rise of an Industry*. New York:

John Wiley and Sons.
Stallen, P.J.M. and Thomas, A. 1988. Public concern about industrial hazards, *Risk Analysis* 8, 237–245.
Stallings, R.A. 1990. Media discourse and the social construction of risk, *Social Problems* 37(1), 80–95.
Starr, C. 1969. Social benefit *vs.* technological risk, *Science* 165, 1232–1238.
Steger, M.E. and Witt, S.L. 1989. Gender differences in environmental orientations: a comparison of publics and activists in Canada and the US, *Western Political Quarterly* 42(2), 627–649.
Sternberg, K. 1991. Hoechst Celanese begins a major emissions reduction, *Chemical Week* 148(12), 9.
Stevens, W.K. 1992. With climate treaty signed, all say they'll do even more, *The New York Times* 13 June, Al.
Stever, D.W. 1989. The use of risk assessment in environmental law, *Columbia Journal of Environmental Law* 14(2), 329–342.
Stigliani, W.M., Doelman, P., Salomons, E., *et al* 1991. Chemical time bombs: predicting the unpredictable, *Environment* 33(4), 4–9, 26–30.
Stobaugh, R. 1988. *Innovation and Competition: the Global Management of Petrochemical Products*. Boston: Harvard Business School Press.
Susman, P., O'Keefe, P. and Wisner, B. 1984. Global disasters: a radical interpretation, in Hewitt K. (ed.), *Interpretations of Calamity*. Boston: Allen and Unwin, 264–283.
Swerdlow, A. 1982. Ladies day at the Capitol: Women Strike for Peace *versus* HUAC, *Feminist Studies* 3, 493–520.
Sun, M. 1989. Market sours on milk hormone, *Science* 246, 876–877.
Teigen, K.H., Brun, W. and Slovic, P. 1988. Societal risks as seen by a Norwegian public, *Journal of Behavioral Decision Making* 1, 111–130.
Thackery, A. and Bowden, M.E. 1989 Uncertain giant: American chemistry on the eve of WWI, *Chemical Week* 2 August, 18–20.
Tiefenbacher, J.P. 1992. Pesticide Drift and the Hazards of Place in San Joaquin County, California. Unpublished Ph.D. dissertation, Department of Geography, Rutgers University, New Brunswick, NJ.
Timmerman, P. 1981. Vulnerability, Resilience and the Collapse of Society. Toronto: Institute for Environmental Studies Environmental Monograph 1.
Titus, A.C. 1986. *Bombs in the Backyard: Atomic Testing and American Politics*. Reno: University of Nevada Press.
Torry, W.I. 1979. Hazards, hazes and holes: a critique of the environment as hazard and general reflections on disaster research, *Canadian Geographer* 23, 368–383.
United Nations Environment Programme (UNEP) 1991. *Environmental Data Report*. Oxford: Basil Blackwell.
—— 1989. *Environmental Data Report*. London: Basil Blackwell.
United States Bureau of the Census 1963. *Census of manufacturing*. Washington DC: Government Printing Office.
—— 1972. *Census of manufacturing*. Washington DC: Government Printing Office.
—— 1982. *Census of manufacturing*. Washington DC: Government Printing Office.
—— 1987. *Census of manufacturing*. Washington DC. Government Printing Office.

US Congress, Office of Technology Assessment 1991. *Complex Cleanup: The Environmental Legacy of Nuclear Weapons Production.* Washington, DC: Government Printing Office, OTA-0-484.

US Department of Transportation 1984. *Emergency Response Guidebook.* Washington DC: Government Printing Office, DOT P5800.3.

—— 1992. Hazardous materials safety. Hazardous Materials Information Service, USDOT, unpublished data.

—— 1980. *Everybody's Problem: Hazardous Waste.* Washington DC: Government Printing Office, SW-826.

US Environmental Protection Agency 1992. Environmental protection—has it been fair, *EPA Journal*, 18(1), 1–62.

—— 1991a. Setting environmental priorities: the rebate about risk, *EPA Journal* 17(2), 1–51.

—— 1991b. *Toxics in the Community: National and Local Perspectives.* Washington DC: Government Printing Office, EPA 560/4–91–014.

—— 1990a. *Toxics in the Community: National and Local Perspectives.* Washington DC: Government Printing Office, EPA 560/4–90–017.

—— 1990b. Reducing Risk: Setting Priorities and Strategies for Environmental Protection. Washington DC: GPO, SAB-EC-90-021.

—— 1989a. *The Toxics-Release Inventory: National and Local Perspectives.* Washington DC: Government Printing Office, EPA 560/4–89–005.

—— 1989b. The greenhouse effect: how it can change our lives, *EPA Journal* 15(1), 1–50.

—— 1988. *Environmental Progress and Challenges: EPA's Update.* Washington DC: Government Printing Office, EPA-230-07-88-033.

—— 1987. *Unfinished Business: A Comparative Assessment of Environmental Problems.* Washington DC: USEPA.

—— 1985. *Chemical Emergency Preparedness Program: Interim Guidance.* Washington DC: Government Printing Office.

US Federal Emergency Management Agency 1981. *Planning Guide and Checklist for Hazardous Materials Contingency Plans.* Washington DC: Government Printing Office, FEMA-10.

US General Accounting Office 1987. *Superfund: Extent of Nation's Potential Hazardous Waste Problem Still Unknown.* Washington DC: Government Printing Office, GAO/RCED-88-44.

US Nuclear Regulatory Commission and US FEMA 1980. *Criteria for Preparation and Evaluation of Radiological Response Plans in Support of Nuclear Power Plants.* Washington DC: Government Printing Office, NUREG-0654, FEMA-REP-1.

US Nuclear Regulatory Commission 1991. *1990 NRC Annual Report.* Washington DC: Government Printing Office, NUREG-1145 volume 7.

Uva, M.D. and Bloom, J. 1989. Exporting pollution: the international waste trade, *Environment* 31(5), 4–5, 43–44.

Verplanken, B. 1989. Beliefs, attitudes, and intentions toward nuclear energy before and after Chernobyl in a longitudinal within-subjects design, *Environment and Behavior* 21, 371–392.

Vlek, C. and Stallen, P.J. 1981. Judging risks and benefits in the small and in the large, *Organizational Behavior and Human Performance* 28, 235–277.

Vonnegut, K. Jr. 1969. *Slaughterhouse Five or the Children's Crusade.* New York: Dell.

Waddell, E. 1977. The hazards of scientism: a review article, *Human Ecology* 5, 69–76.
Wajcman, J. 1991. *Feminism Confronts Technology*. University Park: Pennsylvania State University Press.
Wald, L. 1991. Eastern Europe's reactors don't seem so distant now, *The New York Times* 13 October, 4–4.
Waller, D.C. 1987. *Congress and the Nuclear Freeze: An Inside Look at the Politics of a Mass Movement*. Amherst: University of Massachusetts Press.
Wartenberg, D. and Chess, C. 1992. Risky business: the inexact art of hazard assessment, *The Sciences* (March/April), 17–21.
Wartenberg, D. and Greenberg, M. 1990. Detecting disease clusters: the importance of statistical power, *American Journal of Epidemiology* 132, S156–S166.
Waterstone, M (ed.) 1992. *Risk and Society: The Interaction of Science, Technology, and Public Policy*. Dordrecht: Kluwer Academic.
Watts, M. 1983. On the poverty of theory: natural hazards research in context, in Hewitt, K. (ed.), *Interpretations of Calamity*. Winchester, MA: Allen and Unwin, pp. 231–262.
Whipple, C.G. 1986. Dealing with uncertainty about risk in risk management, in National Academy of Engineering, *Hazards: Technology and Fairness*. Washington: National Academy Press, pp. 44–59.
White, A.L. 1991. Venezuela's organic law: regulating pollution in an industrializing country, *Environment* 33(7), 16–20, 37–42.
White, G.F. (ed.) 1974. *Natural Hazards: Local, National, Global*. New York: Oxford University Press.
White, G.F. 1966. Formation and role of public attitudes, in Jarrett, M. (ed.) *Environmental Quality in a Growing Economy*. Baltimore: Johns Hopkins University Press, pp. 105–127.
White, G.F. 1964. Choice of Adjustment to Floods. Research Paper No. 93. Chicago: University of Chicago Department of Geography.
Whyte, A.V.T. 1986. From hazard perception to human ecology, in Kates R.W. and Burton, I. (eds), *Geography, Resources, and Environment. Volume II Themes from the Work of Gilbert F. White*. Chicago: The University of Chicago Press, pp. 240–271.
Whyte, A.V.T. and Burton, I. 1980. *Environmental Risk Assessment*. SCOPE 15. New York: John Wiley & Sons.
Wildavsky, A. 1979. No risk is the highest risk of all, *American Scientist* 67, 32–37.
Wildavsky, A. and Dake, K. 1990. Theories of risk perception: who fears what and why? *Daedalus* 119(4), 41–60.
Wilhite, D.A. and Easterling, W.E. (eds) 1987. *Planning for Drought*. Boulder: Westview Press.
Wisner, B., O'Keefe, P. and Westlake, K. 1977. Global systems and local disasters: the untapped power of people's science, *Disasters* 1, 47–58.
Withers, J. 1988. *Major Industrial Hazards: Their Appraisal and Control*. New York: Halstead.
Wolpert, J. 1980. The dignity of risk, *Transactions, Institute of British Geographers* 5(4), 391–401.

World Resources Institute 1992. *World Resources 1992–93*. New York: Oxford University Press.
—— 1990. *World Resources 1990–91*. New York: Oxford University Press.
—— 1989. *World Resources 1989*. New York Oxford University Press.
Wright, A. 1990. *The Death of Ramon Gonzalez: The Modern Agricultural Dilemma*. Austin: University of Texas Press.
Yarrow, A.L. 1992. Utilities urged to find nuclear waste sites, *The New York Times* 21 June, 33.
Young, H.F. 1983. *Atlas of United States Foreign Relations*. Washington DC: US Department of State.
Young, R.A. and Kent, A.T. 1985. Using the theory of reasoned action to improve the understanding of recreation behavior, *Journal of Leisure Research* 17(2), 90–106.
Zeckhauser, R.J. and Viscusi, W.K. 1990. Risk within reason, *Science* 248, 559–564.
Zeigler, D.J., Johnson, J.H. Jr. and Brunn, S.D. 1983. *Technological Hazards*. Washington DC: Association of American Geographers Resource Publication.

Index

Complied by Mariana Mossler